For

Serbian and Yugoslav Mauser Rifles

by

Branko Bogdanovic

Illustrated by
Gordana Totic

Translated by
Branka Milosavljevic
Senior Curator, Military Museum
Belgrade, Serbia

Edited by
Dr. Charles H. Cureton
Chief, Museums and Historical Property
U.S. Army Training and Doctrine Command

North Cape Publications®, Inc.

This book is dedicated to the memory of my Father

Acknowledgements

I should like to thank those people who have helped with the preparation of this book. I am grateful to Zastava Arms Factory for giving me the opportunity to work on the factory's history and providing me at the same time with important data on the subject. The translator and I are very much in debt to Dr. Charles H. Cureton for his unselfish assistance and helpful advice. Thanks are also due to Mr. Edmund Parada for providing numerous photographs.

Also, special thanks to the employees of AIM, Middletown, Ohio; Century International Arms, St. Albans, VT; and SAMCO, Miami, FL who provided certain examples of the firearms described in this book for study and photography.

Cover design by Uros Bogdanovic.

Copyright © 2005 by North Cape Publications®, Inc. All rights reserved. Reproduction or translations of any part of this work beyond that permitted by Section 107 or 108 of the 1976 United States Copyright Act without the written permission of the copyright holder is unlawful. Requests for permission or for further information should be addressed to the Permission Department, North Cape Publications.

This publication is designed to provide authoritative and accurate information in regard to the subject matter covered. However, it should be recognized that serial numbers and dates, as well as other information given within, are necessarily limited by the accuracy of source materials.

Information contained in all North Cape Publications®, Inc. material is designed for general historical and educational purposes only. It is not intended to substitute for any owner's manual, safety course, or professional consultation. As such, North Cape Publications, Inc., and all of its affiliates will assume no responsibility or liability for the utilization of such information in printed or electronic form.

ISBN 1-882391-35-7
North Cape Publications®, Inc.,
P.O. Box 1027, Tustin, California 92781

Phone: 800 745-9714; Fax: 714 832-5302.
E-mail: ncape@ix.netcom.com; Website: www.northcapepubs.com

Printed in the USA by Delta Printing Solutions, Valencia, CA 91355

Table of Contents

INTRODUCTION TO SERBIAN AND YUGOSLAV
MAUSER RIFLES AND CARBINES..1
Serbia and Yugoslavia, a Brief History...2
Summary of Serbian/Yugoslav Military History...............................9
Conventions...11
Serbian Cyrillic Alphabet/English Alphabet................................12

CHAPTER 1
SERBIAN MAUSERS, THE MODELS OF 1880 AND 1884....................14
    The Mauser-Milovanovich (Kokinka)
        Model 1880 10.15 mm...14
        Identifying Features,
            the Model 1880 Mauser-Milovanovich............................22
        Model 1880 Markings..24
        The Model 1880 Rifle Deliveries.......................................28
        Model 1880, Military Use..30
    Cavalry and Artillery Carbine Mauser
        Model 1884...31
        Identifying Features, The Model 1884 Mauser
            Cavalry and Artillery Carbines...................................32
        Model 1884 Carbines, Markings..35
        Model 1884 10.15 mm Carbines, Problems..........................35
        Financing the Model 1884 Mauser Carbine..........................38
    Ammunition for the Model 1880/1884,
        the 10.15 x 63 mm R...38
    Burst Cartridge Cases...40

CHAPTER 2
MAUSER RIFLES, M1899, M99/07, M1910, AND CARBINE M1908,
AND DJURICH-MAUSER-KOKA MODEL 1880/07........................42
    Model 1899 Mauser 42
    The Model 1899 Mauser Markings..47
    The Model 1899/07 Rifle and Model 1908 Carbine........................48
        Identifying Features of the Model 1899/07
            Rifle and Model 1908 Carbine.....................................50
    The Serbian Model 1910 Mauser Rifle......................................53
    Identifying Features, Model 1910 Mauser..................................55
        Model 1910 Mauser Rifle Markings....................................60
    The Model 1899, 1899/07, and 1910 After World War I............61
    Summary, Serbian Model 1899, 1899/07 and 1910 Markings............62

Postwar Yugoslavian Model Designations
and Model Markings..63
The Djurich-Mauser-Koka Model 1880/07...............................65
Vindication..68
7 x 57 mm Ammunition..75

CHAPTER 3
TURKISH MAUSERS IN SERBIAN AND YUGOSLAV SERVICE............79
Turkish Model 1887 Rifle And Carbine....................................79
Turkish Model 1890 7.65 x 53 mm Mauser Rifle.........................79
Turkish Model 1893 Mauser Rifle..81
Turkish Model 1903 7.65 x 53 mm Mauser Rifle............................81
Markings...82
Turkish Arms and Serbia..82

CHAPTER 4
THE KINGDOM OF THE SERBS, CROATS, AND SLOVENES: 1918-1929, AND THE KINGDOM of YUGOSLAVIA: 1929-1941............................84
The Model 1924 Mauser Rifle...84
    The Belgian Connection..85
    Model 1924 Rifle and Carbine Details..............................86
    Dimensions..87
    Markings and Finishes..92
The *Chetnik* Model 1924 Assault Rifle.....................................99
    Dimensions..101
    Markings and Finishes..101
    Ammunition and Ammunition Carriers
    for the Model 1924 Mauser..101
Accessories and Bayonet...104
The Sokol Rifle...106

CHAPTER 5
MAUSER AND OTHER RIFLE CONVERSIONS............................111
The Mauser Model 1924 B 7.92 mm Rifle
    (Originally the Mauser 7 mm M1912)...............................112
Mannlicher 7.92 x 57 mm Model 1895M, Model 1895/24
    (Originally, 8 x 50R mm Mannlicher Model 1895)....................115
Mauser 7.92 mm Model 1924C and Model 1924A
    (Originally the 7.92 mm Mauser vz.24)................................125
The Yelen Design..127
Czechoslovakia Selects the Model 1898 Mauser Action.................128

The Czechoslovakian vz.23 Mauser Rifle...........................129
The Czechoslovakian vz.24 Mauser Rifle...........................130
Polish Mausers...132
Mauser Model 1924B 7.92 mm
    (Formerly the Mauser M1898 7.92 mm)........................133
Summary of Mauser Model 1898 Conversions........................135

CHAPTER 6
OCCUPATION...140
The First Partisan's Arms and Ammunition Factory, Uzice.....................143
The *Partizanka* Mauser...144

CHAPTER 7
POST-WORLD WAR II YUGOSLAVIA...149
    The Mauser 7.92 mm Model 1898n...149
    The Mauser 7.92 x 57 mm Model 1924/47...........................153
    The Mauser 7.92 mm Model 1924/52-Č...........................158

CHAPTER 8
THE YUGOSLAV MODEL 1948
    7.92 X 57 MM RIFLE AND ITS VARIATIONS...........................166
    The Mauser 7.92 mm M48—Development and Production..............166
    The Mauser 7.92 mm M48A, M48B and M53...........................167
    A Word about Stock Woods...173
    Export Model Yugoslavian Mausers...182

CHAPTER 9
SNIPER RIFLES...190
The Mauser 7.92 mm M48/52, M1969 and M1993 Sniper Rifles...............190
The Long-Range Sniper Rifle Model 1993
    12.7 mm "Black Arrow"...193

CHAPTER 10
SMALL-CALIBRE RIFLES...201

CHAPTER 11
SPORTING CARBINES...203

Appendix A
Serbian and Yugoslav Military Factories...207

Appendix B
Financing Serbian Mauser Rifles...225

Appendix C
Rifle Markings..231
Appendix D
Serbia's Wars: A Brief History...242
Appendix E
The German Model 1898 Mauser...246
Appendix F
Disassembly/Assembly of Serbian and Yugoslav Mauser Rifles................248
Appendix G
Glossary..257
Appendix H
Bibliography...260

About the Author, Artist, Translator, and Editor..............................270
Books from North Cape Publications®, Inc.....................................272

Tables
Table 1-1, Purchase and Alteration of the Rifles in Serbia from 1881 to 1911..15
Table 1-2, Weapons Having Participated in Competition..........................16
Table 1-3, Technical Data, Model 1880..21
Table 1-4, Markings, Mauser-Milovanoich (Kokinka) Model 1880 10.15 mm..25
Table 1-5, Finishes, M1880 Infantry Rifle, M1884 Carbines.....................26
Table 1-6, Serbian Serial Number Practice..27
Table 1-7, Technical Data, Model 1884 Cavalry and Artillery Carbines........34
Table 1-8, Markings, Model 1884 Cavalry and Artillery Carbines..............37
Table 2-1, Technical Data, Models of 1899, 1899/07, Carbine Model 1908...50
Table 2-2, Finishes, Models of 1899, 1899/07 and Model 1908 Carbine.......52
Table 2-3, Technical Data, Model 1910 Mauser Rifle..............................56
Table 2-4, Markings: The Mauser Rifle M1910, 7 mm..........................57
Table 2-5, Finishes, Model 1910 Mauser Rifle.......................................59
Table 2-6, Technical Data, Model 1880/07 Mauser Rifle........................69
Table 2-7, Markings, Model 1880/07 Mauser Rifle.................................70
Table 2-8, Finishes, M1880/07 Rifle..72
Table 4-1, Technical Data, Model 1924 Mauser Rifle.............................91
Table 4-2, Serial Number Ranges and Markings, Model 1924.................92
Table 4-3, Markings, M1924, M1924CK, Carbine M1924 7.9 mm Rifles.....94
Table 4-4, Finishes, M1924, M1924CK and Carbine M1924 7.9 mm Rifles....98
Table 5-1, Yugoslavian Small-Arms, June 1929....................................111
Table 5-2, Markings, Model 1895 7.9 mm Mannlicher Rifle...................116
Table 5-3, Yugoslavian Small-Arms Inventory, 1937............................136

Table 6-1, German Army Inventory of Serbian Small Arms, 1941.........140
Table 7-1, Rifle Production from 1947 to 1951................................150
Table 7-2, 1013—Rifles and Carbines, Inventory............................150
Table 7-3, Mauser Rifle M24/47 7.92 mm......................................156
Table 8-1, Rifle Production from 1952 to 1967...............................170
Table 8-2, 1013—Rifles and Carbines, April 1967 Inventory..................171
Table 8-3, Technical Data, Post-World War II Yugoslav Rifles..,,,........176
Table 8-4, Markings, Mauser Rifle M98/48, 7.9 mm........................179
Table 8-5, Finishes, M1948, M1948A, M1948B, M24/47,
    M98/48, M24/52-Č, M53 Rifle, and M1969 Sniper Rifle...............181
Table 8-6, Rifles and Carbines, 1962 Inventory..............................186
Table 8-7, Rifles and Carbines, 1970 Inventory..............................187
Table 9-1, Technical Data, Yugoslav Sniper Rifles...........................197
Table 9-2, Markings, Mauser Rifle M1948, M1948A,
    M1948B, M24/52-Č 7.9 mm..199
Table A-1, Factory and Repair Depot Markings and Their Meanings.......223
Tables C-1 through 8: Coat of Arms, Factory, Manufacturers,
    National, Domestic and Export, Various Other Proof Marks
    as of 1970, and Inspectors for Model 1880 through 2005.............234

Yugoslav Partisans fire a captured 50 mm antitank gun at approaching Nazi tanks in December 1943. The gunners are armed with Yugoslav Model 1924 rifles. Photo by George Skrigin.

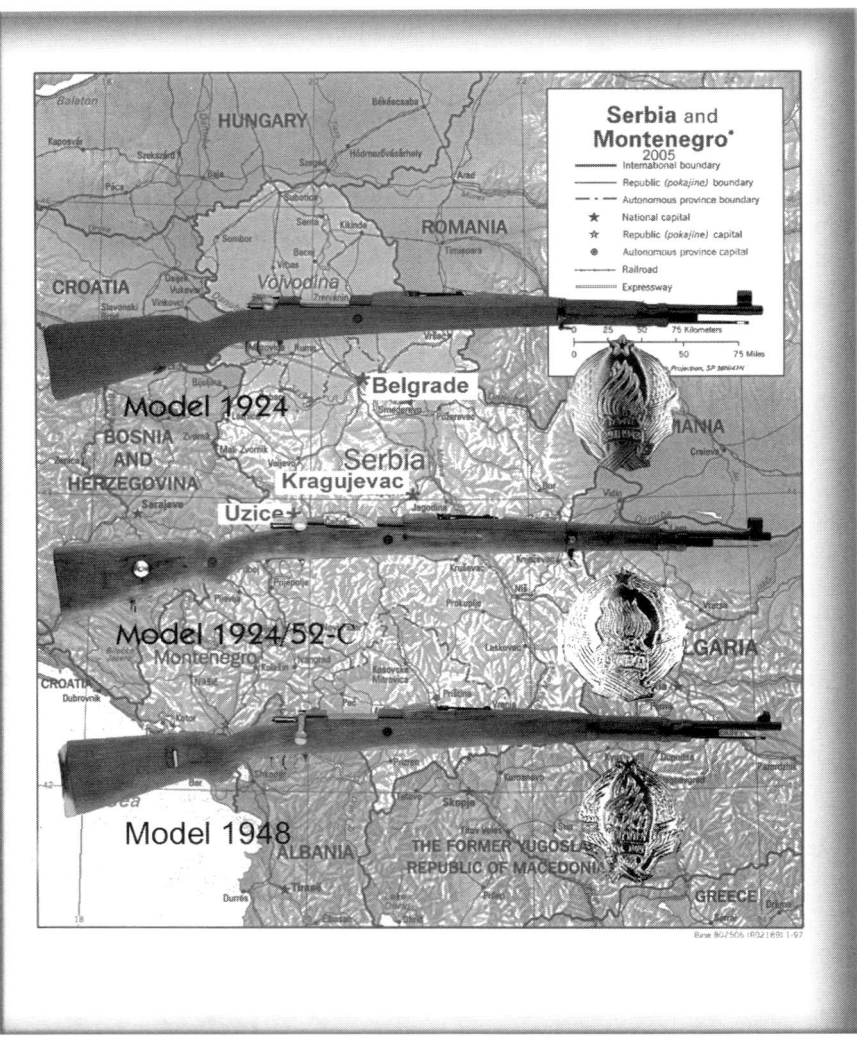

Figure 1. The Mauser bolt-action rifle has armed the military forces of Serbia/Yugoslavia since 1880. Shown superimposed on the map are the Model 1924, the first 7.92 x 57 mm calibre rifle adopted; the Model 1924/52-Č which were built using Czech vz.23 and vz.24 rifles produced during the Nazi Occupation; and the Model 1948, manufactured entirely in Yugoslavia. The Model 1948 rifle can be considered a modernized Model 1924 with features such as the turned-down bolt handle adopted from the K98k. All Serbian/Yugoslav 7.92 x 57 mm Mauser actions are termed "intermediate-ring" Mauser actions as they have the same bolt-head diameter but a slightly shorter action. North Cape Publications collection.

INTRODUCTION
SERBIAN AND YUGOSLAV
MAUSER RIFLES AND CARBINES

In the 1978 movie, *The Deer Hunter*, four of the main characters, Michael (Robert De Niro), Stan (John Cazale), Nick (Christopher Walken) and Axel (Chuck Aspegren), are hunting in the mountains near the American city of Pittsburgh, Pennsylvania. A magnificent buck suddenly appears and Michael kills it with one clean shot. The rifle he used was his favorite, a Mauser Model 70 Zastava made in Yugoslavia. This scene from the cult film about the Vietnam War and its effects on the main characters was fictional; however, the Zastava-made Mauser was real and its use around the world is well documented. Serbian-built Mausers proved to be exceptionally high quality weapons and were used extensively by the Serbian Army from the 1890s well into the last half of the twentieth century. Rifles based on the Mauser system made a substantial contribution to Serbia's campaigns in the Balkan Wars (1912-1913), the First World War (1914-1918), and in Yugoslavia's successful resistance to Nazi aggression and occupation from 1941 to 1945.

Acquiring Mauser weapons involved considerable political and financial difficulties for Serbia. It was a major undertaking that involved many of the country's most prominent politicians and military experts of the time. It also ultimately involved extensive negotiations with the Prussian government concerning selection, contracts, and delivery of the arms. On its own merits, the Mauser rifle was a quality firearm with great appeal for Serbs, but it was the very close connection between many government and ordnance officials with Mauser family members that finally affected the Serbian decision to choose the Mauser rifle.

On 28 April 1884, the president of the Commission responsible for choosing the Serbian weapons, Major Kosta-Koka Milovanovich, married Elisa, daughter of the late Wilhelm Mauser, see Figure 2.

Of no less importance was the fact that the famous Georg Luger, one of the leading gunmakers of the company, Deutsche Waffen- und Munitionsfabriken A.-G (DWM), manufacturers of the Luger "Parabellum" pistol, was part Serbian (through his mother). He was fluent in the Serbian language and unselfishly assisted the Kingdoms of Serbia and Montenegro in choosing rifles and machine guns for their military services, see Figure 3. Until 1914, he provided the Serbian and Montenegrin armies with confidential information concerning arms purchases in Germany by Turkey, Bulgaria, Romania, Greece and Russia.

Serbian and Yugoslav Mausers

Figure 2. Wilhelm Mauser (1834-1882), General Kosta-Koka Milovanovich (1847-1905), and Peter Paul Mauser (1838-1914).

The close ties did not end there. Serbia's purchase of arms and ammunition from DWM and Keller & Co., an arms broker, helped both companies to overcome their financial problems, which in turn helped them develop into giant arms suppliers to the German military.

The first Serbian purchases of Mauser arms were recorded as early as 1880 and continued to the start of World War I in 1914 when because of political circumstances, Serbia and Germany found themselves on opposite sides. In 1928, the Kingdom of Serbs, Croats, and Slovenes (SHS), which became Yugoslavia the following year, began the licensed production of arms based on the Mauser system and continued to do so after the World War II as the new Federal National Republic of Yugoslavia (FNRJ), otherwise known as the Socialist Federal Republic of Yugoslavia (SFRJ). From Mauser military rifles evolved a generation of Yugoslav hunting rifles and carbines which, together with military arms, became one of the country's most profitable exports. Yugoslav Mauser rifles are famous throughout the world. Their importance to both the Serbian and Yugoslav military industry and their world-wide renown and quality make these arms worthy of research and publication.

SERBIA AND YUGOSLAVIA, A BRIEF HISTORY

For much of the 20th century, the lands lying along the Balkan Peninsula in southeastern Europe were known as Yugoslavia, a conglomeration of smaller states that first united in a common nation, named the Kingdom of Serbs, Croats, and Slovenes, in the wake of World War I. In 1929, the nation renamed itself Yugoslavia.

Serbia, the principal state in the Yugoslav Federation, had originally been a vassal principality of the Turkish Empire. The two nations fought

Serbian and Yugoslav Mausers

Figure 3. Georg Luger at Kragujevac in 1909 observing tests of the 7 x 57 mm Maxim M9 machine gun.

sporadic wars between 1389 and 1459 when Serbia was brought under Ottoman rule. Following the Russo-Turkish War (1877) and the Treaty of Berlin, Serbia became an independent kingdom in 1878.

Prince Milan Obrenovich had carefully supported the rebellion of Bosnia and Hercegovina against Turkish rule and in 1876 had declared war on Turkey. The Serbs were unprepared for a major war and were quickly routed by Turkish forces (the first Serbian-Turkish War, 1876). This brought Russia into the war in support of Serbia (during the second Serbian-Turkish war, 1877-1878) and the Turks were forced to ask for an armistice.

The resulting Congress of Berlin (1878) recognized Serbia's complete independence and increased its territory. But Bosnia and Hercegovina were placed under Austro-Hungarian, rather than Serbian administration, a major blow to Serbian aspiration.

Serbia's championship of Pan-Slavism in the Balkans engendered bitter rivalry with both Bulgaria and Austro-Hungary. Prince Milan, who had been proclaimed king in 1882, damaged Serbian prestige by entering an unnecessary and unsuccessful war with Bulgaria in 1885 over

Serbian and Yugoslav Mausers

the question of Eastern Rumelia. Russia supported Bulgaria in this instance and Serbia found herself forced into an uneasy alliance with the Austro-Hungarian Empire. In 1903, the successor to King Milan, King Alexander Obrenovich (ruled 1889-1903), and his queen, both very unpopular, were assassinated, ending the Obrenovich dynasty.

With the accession of King Peter I in 1903, the Karageorgevich dynasty entrenched itself. Peter restored the liberal constitution of 1888 and in 1904 appointed as premier Nikola Pashich, leader of a pro-Russian Radical party. The strengthening of parliamentary government and expansion of the economy greatly raised Serbia's prestige and exerted a powerful attraction on the South Slavs who remained under Austro-Hungarian rule. Austro-Hungary's annexation of Bosnia and Hercegovina in 1908 was designed to quell sentiment in that region for union with Serbia. Aware that under the new geopolitical circumstances they could not enlarge their borders towards the West, the Serbs created a Balkan League (Serbia, Montenegro, Bulgaria, and Greece) dedicated to liberating the Balkan Slavs from Turkish rule.

In 1912 the Balkan League declared war on and defeated Turkey in the First Balkan War. But the League could not agree on division of the spoils. Dissatisfied with its failure to secure a major portion of Macedonia, in 1913, Bulgaria turned against its former allies but was defeated in the Second Balkan War. The successful prosecution of the Second Balkan War resulted in the addition of a large part of Macedonia and all of the province of Kosovo to Serbia, which made her the foremost Slavic power in the Balkans. But it also greatly increased tensions with Austro-Hungary and drove her closer in alliance with Russia and France.

On 28 June 1914, a Serbian nationalist, Gavrilo Princip, a member of the nationalist organization, the "Youth Bosnia," acting without governmental collusion, assassinated Austrian Archduke Franz Ferdinand and his wife, Sophie. On 28 July 1914 the Austro-Hungarian Empire declared war on Serbia. Germany was bound by treaty to support its ally, Austro-Hungary. France, also treaty bound, came in on the side of Serbia, as did the Russian Empire. When Germany marched into Belgium in spite of that small country's declared neutrality, to attack France and outflank the French Army in a bid to capture Paris, the British Empire, which had guaranteed Belgium's independence, was drawn into the war against the Central Powers—Germany and Austro-Hungary. Seeing a chance to regain her lost territories in the Balkans and secure her borders with Russia, Turkey entered the war on the side of Germany and Austro-Hungary late in 1914.

Serbian and Yugoslav Mausers

The Serbian army fought bravely, but in 1915, when Bulgaria joined the Central Powers and Germany reinforced the Austrian army, Serbia was overrun. Serbian troops were evacuated during the winter to Corfu in the Mediterranean Sea after a horrendous retreat through the mountains to the Adriatic coast at San Giovanni de Medua with the loss of more than 20,000 lives. The government had previously withdrawn to Corfu; Serbian, Croatian and Slovenian representatives proclaimed the Union of South Slavs (the Corfu Declaration) on 20 July 1917. In 1918, the Kingdom of the Serbs, Croats, and Slovenes (SHS) officially came into existence in the wake of the collapse of the Austro-Hungarian Empire. Peter I of Serbia was named king of the new nation.

King Peter I Karageorgevich became ill in 1917 and his son, Alexander, served as his regent and succeeded to the throne in 1921 after his father's death. Alexander and Serbia faced demands from Hungary and Bulgaria for revisions to the Treaty of Paris that would grant them Serbian territory. Alexander quickly entered into alliances with Czechoslovakia and Romania to form the Little Entente, in close cooperation but not in alliance with France to resist the Hungarian and Bulgarian demands.

At the same time, relations with their former ally, Italy, became strained over the question of Fiume. Dalmatia had secretly been promised to Italy by the Allies in return for the Italian entry into World War I, and Italian nationalists continued to agitate for the appropriation of the province.

Figure 4. Serb infantrymen on the Salonika front in 1918. The sergeant on the left carries a French M1907/15 Berthier rifle and a Serbian M1899 rifle (repaired at Manufacture d'Armes, St.-Étienne, France—note bent-down bolt handle) with a decorated stock. He is also carrying his M1895 NCO's infantry saber. Photo courtesy of Paul Scarlatta.

Serbian and Yugoslav Mausers

Figure 5. Nazi troops occupy a village in Yugoslavia in April 1941. Surrendered Yugoslav vehicles line either side of the road.

Internal problems plagued Alexander as well. Because of nationalistic tensions and the strong and dangerous activity of the Communist Party, King Alexander dissolved the parliament, changed the name of the country to Yugoslavia (Jugoslavija) in 1929 and declared a dictatorship which did not end until 1931. Croatia and now Macedonia as well, supported by the Bulgarian nationalists, continued to agitate and in 1934, King Alexander was assassinated in Marseilles, France.

His son, Peter II, became regent under his father's cousin, Prince Paul. Prince Paul negotiated with the Axis powers, Germany and Italy, to maintain independence and neutrality. Yugoslavia was pregnant with difficulties. Neither national, nor economic and peasant issues could be resolved. In 1939, Serbs and Croats agreed on wider autonomy for Croatia; a new Yugoslav government was formed with the leader of the Croatian Peasant Party, Vlatko Macek, as the vice premier.

In March 1941, under intense pressure from Nazi Germany, Yugoslavia signed the Axis Tripartite Pact. But the opposition led a bloodless military coup which ousted Prince Paul as regent and declared a policy of neutrality. The action so angered Hitler that in April 1941 he sent German troops, reinforced by Bulgarian, Hungarian, and Italian armed forces, to invade Yugoslavia, thus delaying his attack on the Soviet Union by a fateful six weeks. The country was occupied within eleven

Serbian and Yugoslav Mausers

days. A German puppet state was created in Croatia, led by Ante Pavelich, head of the *Ustase* which later became a Croatian terrorist organization. Dalmatia, Montenegro, and Slovenia were divided between Italy, Hungary, and Germany. Serbian Macedonia became part of Bulgaria. Serbia became a German occupation zone.

King Peter II established a government in exile in London, and many units of the Yugoslav army fled into the mountains to conduct guerrilla warfare against Axis troops. Two major resistance groups developed. One was the nationalist *Chetniks* under Dragoljub-Draza Mihajlovic and the second was the Communist Partisans led by Josip Broz Tito. At first, the two groups maintained a distant cooperation but by 1943, an open civil war had developed between the *Chetniks* and Partisans. Tito was supported by the USSR and by Great Britain which provided extensive military aid and advisers. Without the support of the Allies, primarily Great Britain, the *Chetniks* became more and more hesitant in their operations against Axis occupation troops. As a result, King Peter transferred the military command of the resistance to Tito.

Figure 6. Josip Broz Tito led the Communist-dominated Partisan guerrilla war against the Nazis during World War II.

In late October 1944, the Germans were driven from Yugoslavia by a combined offensive of the Partisans and the Soviet Red Army. Tito's national liberation council merged in November with the royal government. By March 1945, Tito was elected premier. The non-Communist members of the government resigned and were arrested. In November 1945, national elections gave the government to Tito by default as the opposition abstained from the elections. A constituent assembly was formed

Serbian and Yugoslav Mausers

Figure 7. A Partisan prepares a demolition charge to destroy a bridge in 1944.

and proclaimed a federal people's republic.

The constitution promulgated in 1946 provided autonomy to six newly created republics within Yugoslavia, but with the actual power remaining in the Communist party. Communism was imposed harshly within Yugoslavia and any opposition was smashed. Until 1948, a close alliance was maintained with the Soviet Union, when disagreements over the direction and activities of the Cominform caused the Soviet Union to expel Yugoslavia.

From that point on, Tito's government pursued an ostensibly neutral foreign policy and even received economic and military assistance from the West. In 1953, Yugoslavia abandoned strict communism and the collectivization of agriculture. A defense pact with Greece and Turkey, outside the purview of NATO, was concluded in 1954. Relations with the Soviet Union improved after the death of Stalin but the Soviet occupation of Hungary in 1956 and Czechoslovakia in 1968 created additional strains.

Under Tito, Yugoslavians enjoyed greater freedom than other Eastern European countries in the Warsaw Pact which Yugoslavia declined to join. Tito's style of neutrality was to walk a fine line "between East and West," creating a "Third World" instead that maintained its distance from both superpowers.

The six republics and two autonomous provinces of Serbia within Yugoslavia gained additional autonomy during the 1970s as the economy stagnated. Tito died in 1980 and was succeeded by an unsuccessful collective leadership who were unable to hold together the disparate peoples of Yugoslavia. The country began to slowly disintegrate. Economic problems and ethnic divisions continued to deepen in the 1980s, and the foreign debt grew significantly.

In January 1990, the Communist Party ended its majority role in the government. Croatia and Slovenia declared independence in June

Serbian and Yugoslav Mausers

1991. In April 1992, Serbia and Montenegro established a new Federal Republic of Yugoslavia. Fighting raged between Serbia/Montenegro, Croatia and Slovenia, between Serbs and Moslems in Bosnia as well as fighting between Serbs and the Albanian rebels in Kosovo. The fighting was ended in June 1999 when a 50,000-member multinational force was sent to Kosovo. In February 2003 Serbia and Montenegro established a new political entity known as the Federation of Serbia and Montenegro.

SUMMARY OF SERBIAN/YUGOSLAV MILITARY HISTORY

Serbia became a vassal principality of the Ottoman Empire in 1389 but engaged in sporadic wars until Turkey occupied Serbia in 1459.

Following two uprisings against Turkish rule in 1804 and 1815, the Principality of Serbia achieved home-rule status but continued to recognize Turkish dominion.

Following the war with Turkey (1876-1878), which saw Russia intervene on behalf of Serbia, the Congress of Berlin granted Serbia full independence in 1878. In 1882, the Kingdom of Serbia was established.

Serbia and Bulgaria fought a war in 1885 for territorial acquisition and political domination over the Balkans.

The War of the Customs was primarily a political-economic conflict with the Austro-Hungarian Empire that lasted, on and off, from 1906 to 1911. It resulted from the Austro-Hungarian Empire's attempt to impose an economic embargo on Serbia for breaking trade agreements. Among other things, it forbade the import of livestock and agricultural products from Serbia.

The First Balkan War was fought in 1912. Serbia, Bulgaria, Greece and Montenegro allied themselves to expel the Turks from the Balkans.

The Second Balkan War was fought in 1913. Serbia, Greece, Montenegro, and Romania (who had entered the war late) fought Bulgaria over the division of conquered Turkish territories.

World War I (1914-1918) began ostensibly with the assassination of the Austrian Archduke Ferdinand, and his wife, Sophie, by a Serbian,

Serbian and Yugoslav Mausers

Gavrilo Princip. The Austro-Hungarian Empire demanded Serbian territory in reparations. Serbia refused and war was declared. Germany came in on the side of Austro-Hungary, Russia on the side of Serbia, France on the side of Russia, and Turkey on the side of Germany. The British Empire had guaranteed Belgium's neutrality and entered the conflict when Germany invaded Belgium.

As the war was ending, the Kingdom of the Serbs, Croats and Slovenes (SHS) was established in 1918. In 1929, the country was renamed Yugoslavia.

World War II (1939-1945) saw Yugoslavia invaded by Nazi Germany, Hungary, Italy and Bulgaria. Guerrilla warfare against the Germans was begun by Serbian and Croatian resistance groups.

King Peter I established a government in exile in London and in 1943, threw his support to Tito's Partisans.

German forces were expelled from Yugoslavia in October 1944 and the Yugoslav state was established with Tito as its head.

Communism was imposed in 1946 with the collectivization of agriculture and the nationalization of industry.

Subversive activities by the Cominform led Tito to break with the Soviet Union in 1948 and declare Yugoslavia's neutrality.

Collectivization of agriculture and strict communism was abandoned in 1953. Tito died in 1980 and in January 1990, the Communist Party ended its monopoly on power.

Between 1992 and 1999, civil war raged in the former Yugoslavia. It was ended by an international military force.

In 2003, Serbia and Montenegro established a new national entity, the Federation of Serbia and Montenegro.

Serbian and Yugoslav Mausers

Conventions

Between 1880 and 1948, Serbia was known by four names: 1880 to 1918 as Serbia, 1918 to 1929 as the Kingdom of the Serbs, Croats and Slovenes (SHS), from 1929 to 1941 as the Kingdom of Yugoslavia (sometimes spelled Jugoslavia), and as the Federal Republic of Yugoslavia after 1945. In this text, the author has used the appropriate name of the country for the period of time being discussed.

Markings on rifles described in the text are enclosed in quotation marks. The quotation marks are not part of the marking, unless so noted.

When measurements are given from a point to a hole or other opening, that measurement is understood to be to the center of the hole or opening, unless otherwise noted.

All reference directions are given from the shooter's standpoint while looking toward the muzzle with the rifle shouldered. Thus the right side refers to the side with the bolt handle, etc.

Artists' drawings are used throughout this text in preference to photographs for clarity and to emphasize certain aspects.

The Mauser and other rifles described in this text were manufactured using the metric system of measurement. All dimensions are given according to the metric system followed by the same dimension according to the English system using decimal inches. Millimeters are converted to decimal inches by dividing by 25.4; decimal inches are converted into millimeters by multiplying by 25.4.

The usual practice in Serbia/SHS/Yugoslavia was to mark the "model" designation of a rifle after the year the model was adopted (1880, 1924, or 1948, for instance) as M80, M24 and M48.
Major variations of that model, or if the rifle underwent a major rebuild, were indicated by adding the last two digits of the year the rebuild began, i.e., Model 1924/47.
Changes to the model of a lesser nature were denoted by the addition of a sequential letter to the model year, i.e., Model 1948A, Model 1948B. The letters were, of course, marked on the rifle in the Cyrillic alphabet. These letters were sometime separated by a space from the

Serbian and Yugoslav Mausers

Model year, and sometimes not. When quoting the marks directly from the rifle, the marking will be spaced exactly as on the rifle.

See below for a comparison of the Cyrillic and Latin alphabets.

To avoid confusion with other nations' Mauser rifle model markings, each model will be referred to in the text by its full model year designation, calibre and type, i.e., Model 1880 10.15 mm Mauser Rifle, Model 1924 7.92 x 57 mm Mauser Rifle, Model 1948A 7.92 x 57 mm Mauser Rifle. If space does not allow or brevity is desired, the designation may be shortened as follows, Model 1924/47 or Model 1948A.

The European practice is to use a comma in place of a decimal point, i.e., 7,9 rather than 7.9. In the text we have substituted decimal points which are more familiar to North American readers. But the comma instead of the decimal point may be used in tables and illustrations to preserve the flavor of the author's writing.

SERBIAN CYRILLIC ALPHABET/ENGLISH ALPHABET

А	A as in father
Б	B
В	V
Г	G as in go
Д	D
Ђ	Almost like the j in jack
Е	E as in end
Ж	S as in measure
З	Z
И	EE as in seen
J	Y as in you
К	K
Л	L
Љ	lli as in million
М	M
Н	N
Њ	Ny as in canyon
О	O

Serbian and Yugoslav Mausers

П	P
Р	R
С	S
Т	T
Ћ	Almost as ch in church
У	OO as in moon
Ф	F
Х	ch as in Bach*
Ц	ts as in cats
Ч	Ch as in church
Ш	Almost like j as in jeep
Ш	Sh as in sharp

* The Cyrillic letter X is also used in this book in transliterations with the English letter "H" (as seen in the TEX=TEH, CXC=SHS)

NOTE: Some Cyrillic letters may be retained in transliteration in this text; i.e., *J-Y* in some cases, such as "Vo*j*no." Also, certain words may retain Serbian spelling rather than American/foreign standardized spellings.

CHAPTER 1
SERBIAN MAUSERS,
THE MODELS OF 1880 AND 1884

During the 1880s, the Serbian government carefully studied the modernization of the Serbian army. Weapons standardization was a significant part of army reforms and the objective was to find firearms that could meet present and future needs. Serbia lacked sufficient funds for experimentation and this forced the army to choose wisely.

In 1878, the Treaty of Berlin ended the Russo-Turkish War and the Principality of Serbia became independent. In 1882, the Kingdom of Serbia was established. Between 1881 and 1911, Serbia acquired nearly 400,000 rifles and carbines from foreign suppliers to equip its military and police. This was an immense burden for so small a country and huge amounts of time and energy were devoted to raising the needed loans to equip their military force. Not until after the turn of the last century, did Serbia develop the infrastructure to manufacture its own military weapons. And then, its capacity was limited until after World War I when Serbia was joined with Croatia, Bosnia, Dalmatia, Hercegovina, Vojvodina, Montenegro, and Slovenia to form the Kingdom of the Serbs, Croats and Slovenes. In 1929, the country was renamed Yugoslavia.

Table 1-1 provides a listing of rifle and carbines procured by Serbia between 1880 and 1911.

THE MAUSER-MILOVANOVICH (KOKINKA) MODEL 1880 10.15 MM
On 6 January 1879, the Minister of War, Yovan Miskovich, organized the Artillery Committee headed by Colonel Sava Grujich and ordered them to develop a methodology to select effective firearms of lasting quality. Sava Grujich's organization developed the criteria for the "Commission to Choose a New Rifle Model." The initial members of the commission were Major Kosta Milovanovich (President), Captain Pavle Najdanovich, Captain Kosta Kostich, Captain Pavle Jurisich-Sturm, Lieutenant Lazar Petrovich, Lieutenant Yovan Pavlovich and the civilian supervisor of the Gunsmith Shop, Kragujevac, Mr. Vasilije Vasikich. The commission hoped to attract both domestic and international manufacturers, and announced a competition in which all except mediators and arms traders could take part. Out of all the designs submitted, only twenty-nine different weapons systems from seventeen firms were considered worthy of consideration. In all, the commission chose forty different rifles for further evaluation, see Table 1-2.

Serbian and Yugoslav Mausers

Table 1-1
Purchase and Alteration of the Rifles in Serbia from 1881 to 1911

Year	M1880 10.15 mm Infantry Rifle	M1880 10.15 mm Sporting Rifle	M1880 10.15 mm Barrel	M1880 Spare parts	M1884 10.15 mm Cavalry Carbine	M1884 10.15 mm Artillery Carbine	M1899 7 mm Infantry Rifle	M1899/07 7 mm Infantry Rifle	M1899/07 7 mm Barrel	M1899/07 7 mm Cavalry Carbine	M1908 7 mm Cavalry Carbine	M1910 7 mm Infantry Rifle	M1880/07 7 mm Infantry Rifle
4 August 1881– 27 February 1884	100,000												
1884		20,000											
17 October 1884– 28 October 1885			1,000	125,294									
May 1899– December 1900					4000	4000							
December 1900							90,000						
December 1907–June 1908								34,000	50,900	10,800			
October 1910– June 1911												32,000	
1907– 4 March 1911													43,000

15

Serbian and Yugoslav Mausers

Table 1-2 Weapons Having Participated in Competition		
Manufactured by	System	Model 1898 & System
SIG, Schweizerische Industrie-Gesellschaft, Neuhausenbei Schaffhausen (Schweiz)	Feterli jednometka za svajcarski fisek (metak)	10.4 mm Single Shot Cartridge Rifle M1868, Vetterli
	Feterli jednometka za Martinijev fisek	.45 Single Shot Cartridge Rifle M1868, Vetterli
	Feterli Magacinka (repetirka B visemetka) za svajcarski fisek	10.4 mm Repeating Rifle M1879/81, Vetterli
	Karabin Feterli jednometan za svajcarski fisek	10.4 mm Single Shot Cartridge Carbine M1868, Vetterli
	Karabin Feterli Magacinka za svajcarski fisek	10.4 mm Repeating Carbine M1879/81, Vetterli
Gusstahl und Waffen-Fabrik Witten, vormals Bergen et Comp. Witten a.d., Ruhr, Germany (Nemacka)	Martini B Henri (rumunjski)	11 mm Rumanian Single Shot Cartridge Rifle M1879, Martini-Henry "Perfestione"
Thomas Sederl Wien, Austria (Oesterreich-Ungarn)	Pibodovaca srpska	14.9 mm Serbian Single Shot Cartridge Rifle M1870, Peabody
	Valmisbergova magacinka	11 mm Single Shot Cartridge Rifle M1881, Valmisberg
C. D. Haenel Militair et Luxus-Waffenfabrik. Suhl Thuringen, Germany (Nemacka)	Mauzer od 1871. godine	11 mm Single Shot Cartridge Rifle M1871, Mauser
	Timier B Vert.	11 mm Single Shot Cartridge Rifle M1880, Timmer

Serbian and Yugoslav Mausers

Table 1-2, cont.
Weapons Having Participated in Competition

Manufactured by	System	Model 1898 & System
C. D. Haenel, cont. Militair et Luxus-Waffenfabrik. Suhl Thuringen, Germany (Nemacka)	Bornmiler	11 mm Single Shot Cartridge Rifle M1870, Bornmueller
	Karabin Mauser od 1871. godine	11 mm Single Shot Cartridge Carbine M1871, Mauser
Manufactur des armes à feu August Francotte, Liége, Belgium (Belgique)	Martini-Frankot	.45 Single Shot Cartridge Rifle-Martini-Francotte
William Soper. Reading, England (Ingleska)	Soper	.455 Single Shot Cartridge Rifle M1868, Soper
Winchester Repeating Arms Company, New Haven, Conn., U. S. A. (America)	Hockisova sa amerikanskim fisekom	.45 Repeating Rifle M1878, Hotchkiss
Schmidt Oberstl. Director der eidgen. Waffenfabrik Bern (Schweiz)	Smit-Miller	10.4 mm Single Shot Cartridge Rifle M1879 B Schmidt-Miller
The Lee Arms Company Bridgeport, Conn., U. S. A. (America)	Li magacinka	.45 Lee Repeating Rifle M1879
Gebrueder Mauser & Comp. Oberndorf a. Neckar (Wuerttemberg)	Mauzer od 1878. godine	11 mm Single Shot Cartridge Rifle M1878, Mauser
	Mauzer-Koka s'paralelnim zlebovima	10.15 mm Serbian Single Shot Cartridge Rifle M1880, Mauser-Koka
	Mauzer-Koka s'klinastim zlebovima	10.15 mm Serbian Single Shot Cartridge Rifle M1880 B, Mauser-Koka
	Mauzerov karabin od 1878. godine	11 mm Single Shot Cartridge Carbine M1878, Mauser

Serbian and Yugoslav Mausers

Table 1-2, cont. Weapons Having Participated in Competition		
Manufactured by	**System**	**Model 1898 & System**
Small Arms and Metal Cartridge Company Limited, Birmingham, England (Ingleska)	Flid-Henri	.455 Single Shot Cartridge Rifle B Field-Henry
	Svinbern-Henri	.455 Single Shot Cartridge Rifle B Swinburn-Henry
	Martini-Henri	.455 Single Shot Cartridge Rifle M1871, Martini-Henry
The National Arms & Ammunition Company Limited, Birmingham, England (Ingleska)	Delji-Edz	.45 Single Shot Cartridge Rifle M1879 B Deeley-Edge
	Mauzer od 1871. god.	11 mm Single Shot Cartridge Rifle M1871-Mauser
Misonne et Schlesser. Charleroi (Belgium)	Komblen	11 mm Single Shot Cartridge Rifle M1871-Comblain
N. v. Drayse Sommerda, Germany (Nemacka)	Drajze sa uzduznim zatvaracem	15.43 mm Single Shot Cartridge Rifle M1865 Dreyse (bolt action)
	Drajze sa poprecnim zatvaracem	15.43 mm Single Shot Cartridge Rifle Dreyse (breech-block)
Milisav Petrovic, puskar Krajinske brigade, Negotin (Serbia)	Sumadinka	14.9 mm Serbian Single Shot Cartridge Rifle M1870/79B Milisav Petrovich-"Sumadinka"
Josif Lambaher, majstor vojne fabrike, Kragujevac (Serbia)	Lambaher	14. 9 mm Serbian Single Shot Cartridge Rifle M1878-Lambacher

Serbian and Yugoslav Mausers

Table 1-2, cont. Weapons Having Participated in Competition		
Manufactured by	System	Model 1898 & System
Oesterreichische Waffen-Fabriks B Gesellschaft (Werndl) Steyr, Austria (Ober-Oesterreich)	Mauzer od 1871. god.	11 mm Single Shot Cartridge Rifle M1871, Mauser
	Martini Henri za austrijski fisek	11 mm Single Shot Cartridge Rifle M1871 B Henry-Martini
	Gra od 1874. godine	11 mm Single Shot Cartridge Rifle M1874-Gras
	Kropacek magacinka za austrijski fisek	11 mm Repeating Rifle M1878, Kropatschek
	Kropacek magacinka za svedski fisek	10.15 mm Repeating Rifle M1878, Kropatschek
	Magacinka Verndl sa 3 magacinka za austrijski fisek	11 mm Repeating Rifle, Werndl
	Spitaljski za austrijski fisek	11 mm Single Shot Cartridge Rifle, Spitálský
	Martini-Verndl za austrijski fisek	11 mm Single Shot Cartridge Rifle B Martini-Werndl

The ballistics, technical performance, and field-testing of these weapons took place at the Banjica firing range in Belgrade. These tests reduced the number of weapons considerably and the original German Mauser Model 1871 and its improved variant the Mauser Model 1878 were the two weapons selected for further evaluation. From his single-shot bolt-action rifle Model 1871, Wilhelm Mauser developed a new repeating rifle designated the Model 71/78. It was patented under DRP No. 15,204, dated 23 January 1880.

The new rifle was first shown and tested at the Banjica range. Major Kosta Milovanovich was impressed but wanted to improve its ballistics. Consequently, he gave up his position as President of the Commission to Major Pavle Shafaric and devoted himself totally to redesigning the

Serbian and Yugoslav Mausers

Mauser barrel. He adopted the 10.15 mm calibre Jarman cartridge used by Norway and experimented with two kinds of rifling, the classical parallel and the modern wedge. He collaborated with the Gebrueder Mauser and Company factory's Obermeister, August Gaiser (1835-1911). By March 1880, their single-shot version made it to the final trials along with Wilhelm Mauser's Model 71/78, a Kropatschek repeating rifle from OEWG Steyr, and both versions of the Mauser-Koka, see Figure 1-1.

NOTE: The lands in "wedge" rifling taper or narrow from breech to muzzle. In the case of the Serbian Model 1880 rifle 10.15, the lands were 0.181 inch (4.6 mm) wide at the breech end and 0.157 inch (4 mm) at the muzzle.

The Milovanovich rifle using wedge rifling received very positive reviews in the Commission's report to the Ordnance Board dated 26 August 1880. Their only criticism concerned the rear sight, which was graduated to 300 steps (218.8 yards or 200.1 meters) with the leaf raised. A modified sight resulted that was graduated to 400 steps (291.8 yards or 266.8 meters) with a leaf sight that was scaled from 500 to 2,500 steps (364.7 yards or 333.5 meters to 1,823.6 yards or 1,667.5 meters) with the notch on the leaf top allowing for aimed fire up to 2,700 steps (1,969.5 yards or 1,800.9 meters). The Serbs measured distances in steps at this time. One step equaled 0.7294 yard or 0.667 meter.

At last, on 21 October the Commission voted 10 to 6 to recommend the adoption of the single-shot breechloader, Model 1880 Mauser-Milovanovich with wedge rifling in 10.15 calibre for the Serbian Army, refer to Table 1-2 and see Table 1-3. Wilhelm Mauser, however, stated in his letters that there were 12 to 4 votes for his design.

Fig. 1-1a. The Model 1880 10.15 mm Mauser-Milovanovich M1880 infantry rifle was a refinement of the German Model 1871 11 x 60 mm rifle.

Serbian and Yugoslav Mausers

Figure 1-1b. Top, Model 1880 10.15 mm Mauser-Milovanovich M1880 infantry rifle. Bottom, a Model 1880 presentation rifle.

\multicolumn{3}{c}{Table 1-3 Technical Data, Model 1880}		
Model	\multicolumn{2}{l}{Model 1880 Serbian Mauser Infantry Rifle}	
Calibre (mm)	10.15	
Rifling	wedge, 4 grooves, 0.25 mm (0.0098 inch) deep and from 4.6 to 4.0 mm (0.18 to 0.15 inch) wide breech to muzzle; 1 turn in 550 mm, right hand (pitch of 31 19' 5")	
Magazine	none - single-shot only	
Loading system	manual insertion of a cartridge in chamber or bolt-way	
Length overall	mm	1,295
	inch	51
Barrel length	mm	801
	inch	31.53
Weight	gm	4,500
	lbs	9.9
Sights	Steps (yards/meters)	(front) unprotected barleycorn; (rear) fixed "battle" sight for 400 steps (291.8 yards or 266.8 meters) and a leaf with an elevation slide extension for 500 to 2700 steps (364.7 to 1,969.5 yds or 333.5 to 1800.9 m)
Action	Small-ring Mauser action. Receiver ring is 1.32 inches (33.5 mm) long with a screw spacing of 9.53 inches (242.1 mm) and a bolt body 6.53 inches (165.9 mm) long.	

Serbian and Yugoslav Mausers

Identifying Features, the Model 1880 Mauser-Milovanovich

The Model 1880 10.15 mm Mauser rifle approved for the Serbian Army was basically a redesigned Model 1871 German Mauser, single-shot infantry rifle, see Figure 1-2.

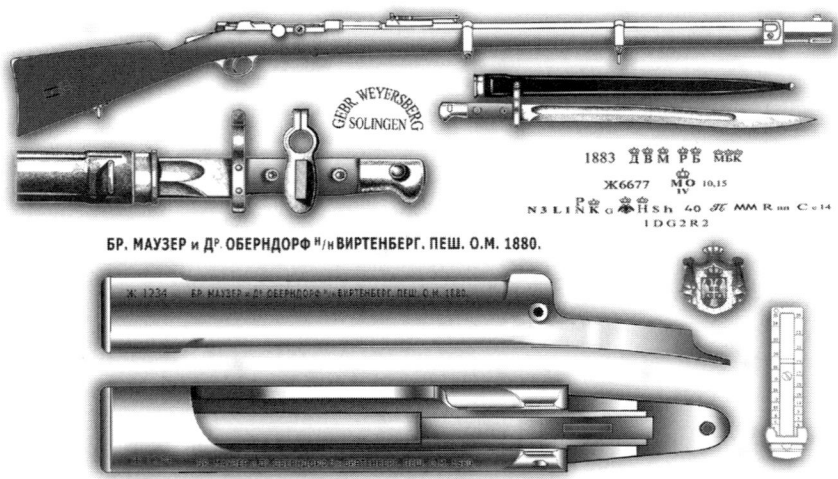

Figure 1-2. The Mauser-Milovanovich Model 1880 10.15 mm Rifle with bayonet.

The Serbian Model 1880 Mauser can easily be distinguished from the German Model 1871 Mauser: 1) the trigger guard lacked the reinforcement behind the trigger guard bow, 2) the barrel bands were the clamping type without barrel band springs inletted into the forend, 3) the forward barrel band was a true band rather than a nose cap as in the German rifle and the forend did not protrude past the band, 4) the bayonet mount was centered on the right side of the forward barrel band which was secured with a screw that entered from the right side of the band, penetrated the forend and threaded into the left side of the band, 5) the rear sling swivel was mounted on the bottom of the buttstock rather than on the trigger guard as on the German rifle, and 5) the rear tang behind the receiver was raised above the line of the wrist to serve as a bolt guide, similar to that of the Italian Model 1870 Vetterli rifle. 6) The bolt handle was straight and protruded to the right at a ninety-degree angle. It ended in a ball 0.862 inch or 21.9 mm in diameter. A large screw held a washer, or retaining ring, which prevented the bolt from being accidentally drawn out of the bolt race. To remove the bolt, the screw and retaining ring had

Serbian and Yugoslav Mausers

to be removed first. The bolt had a short, clawed extractor mounted on a ring inset in the bolt body. The safety lever was mounted on an axle at the rear of the bolt body, see Figure 1-3, arrow.

Figure 1-3. Model 1880 bolt assembly.

NOTE: The rifle's popular names, "Kokinka" and "Koka," derive from Major Kosta Milovanovich's nickname.

The Model 1880 rifle rear sight was graduated in steps. The rear sight base was graduated to 300 steps (218.8 yards or 200.1 meters) while the sight leaf was graduated from 500 to 2,700 steps (364.7 yards or 333.5 meters to 1,969.5 yards or 1,800.9 meters).

Under the supervision of Milutin Markovich, appointed as a controller on 1 December 1881, the barrels were manufactured by "Simson & Company" of Suhl, Germany. The steel came from Asthever in Ahlen, Westphalia, Germany. The springs were produced in the well-known Austrian factory Gebrueder Boehler & Company AG (Edelstahlwereke, Wien I, Elisabethenstrasse 12 Wien).

The wood for the full-length walnut stock was felled between 1871 and 1873 in Pfalz in the Rheinland. The steel furniture and the barrels were rust-blued in a steam cabinet; the deep blue obtained was the result of seven steam-cabinet treatments.

The Serbian Model 1880 was somewhat longer and heavier than the German Model 1871. It was 51 inches (1,295 mm) long overall compared to 48.7 inches (1,237 mm) for the German Model 1871 rifle. The barrel was 31.53 inches (800.1 mm) long compared to 29.44 inches (747.8 mm) and the Serbian rifle weighed 9.9 lbs (4.49 kg) compared to 9.0 lbs (4.1 kg) for the German rifle.

The Model 1880 rifle was equipped with a yataghan-type bayonet. Its blade differed from the straight, knife blade of the German bayonet, refer to Figure 1-2. The bayonet was manufactured by Gebrueder Weyersberg in Solingen-Hohscheide, Germany. Bayonet production was supervised

Serbian and Yugoslav Mausers

by Milisav Petrovich, a specialist sent to the Gebrueder Weyersberg factory from Kragujevac. The scabbard was made of wood and covered with black leather. A staple mounted on the left side protruded through the leather frog worn on the infantryman's belt and was secured with a leather tongue.

Model 1880 Markings
The manufacturer's markings were stamped with a roll die on the left side of the receiver in Serbian (Cyrillic characters), followed by the model number in Arabic numerals (1880). Refer to Figure 1-2 and see Figure 1-4. The receiver markings translate to "Mauser Brothers & Co. Oberndorf

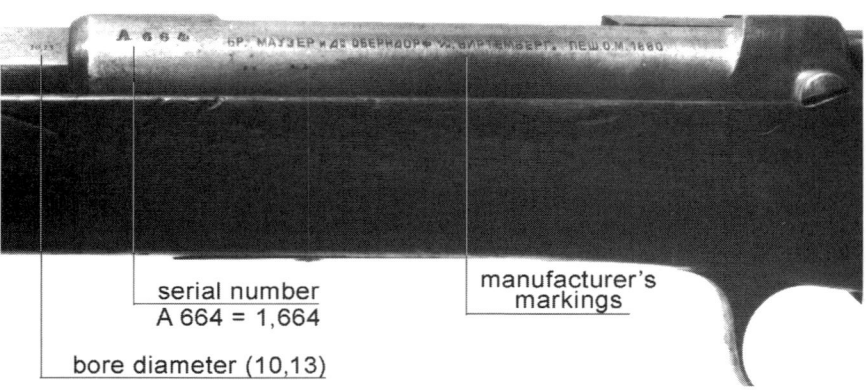

Figure 1-4. Model 1880 receiver and bolt markings.

a/N, Wuerttemberg, Infantry-Breechloading Model 1880." Each rifle was also stamped with the monogram of Prince Milan Obrenovich IV, "MO IV" below "crown/date/calibre/proof marks" and a serial number of up to five figures preceded by a Cyrillic letter.

Instruction AN 1743, dated 16 July 1883, directed that starting with the "G" series (rifle, serial number 30,001), the rifles would also carry a crown proof mark. Barrels whose bore diameters were larger than specified were also marked "X" next to the calibre marking.

Every major part was marked with an inspector's mark (see Table 1-4 and refer to Figure 1-2, Figure 1-4, and Appendix C-6 for examples).

Serbian and Yugoslav Mausers

Table 1-4
Markings, Mauser-Milovanovich (Kokinka) Model 1880 10.15 mm

Marking	Barrel: Receiver Ring	Barrel: Top	Barrel: Left	Barrel: Right	Receiver: Left	Receiver: Right	Bolt Assembly	Stock	Trigger Assembly	Trigger Guard Plate
Coat of Arms and Model Marking					x*					
Manufacturer's Markings					×					
Country Name										
Year of Production							×			
Ruler's Monogram	×									
Full Serial Number				×				×		
Last Three or Four Digits of Serial Number	×						×		×	×
Proof Mark	×			×		×				
Inspection Mark	×			×		×	×	×		
Acceptance Mark	×			×		×				
Calibre Mark	×									

X* Large "X" is without Coat of Arms

Serbian and Yugoslav Mausers

The right side of the buttstock was stamped with a crown over the letter Cyrillic "H" (Latin "N").

The finish applied to each part is shown in Table 1-5.

Table 1-5 Finishes, M1880 Infantry Rifle, M1884 Carbines	
Part	Finish
All wood parts	Sanded smooth and oiled
Barrel	Blued, seven steam cabinet treatments
Receiver	Blued, seven steam cabinet treatments
Front Band, Middle Band, Lower Band	Blued, seven steam cabinet treatments
Metal plate on the underside of the butt	Blued, seven steam cabinet treatments
Swivels	Blued
Butt Plate	Blued, seven steam cabinet treatments
Front Sight Blade	Blued by heating
Rear Sight Base	Blued, seven steam cabinet treatments
Sight Leaf	Blued
Slide	Blued
Slide Catch	Blued
Screws	Finished with a black coating
Bolt Parts	Finished in-the-white
Bolt Screw with Washer	Blued
Trigger Guard, Trigger Guard Plate	Blued, seven steam cabinet treatments

The serial number range for the Model 1880 Mauser-Milovanovich rifle was A1 through J10,000 (110,000). The Serbs followed the German system of serial numbering (also used by many other Continental European nations). In this system, the first series of 10,000 rifles was designated by the letter "A," the second series of 10,000 by the letter "B," and so on to the end of production. This prevented the serial number from becoming

Serbian and Yugoslav Mausers

unduly long. Serbia followed the same program but substituted letters from the Cyrillic alphabet, see Table 1-6. All major parts of the rifle carried the full or partial serial number stamped on the receiver. If the serial number on a part does not agree with that on the receiver, it was probably replaced at one time or another. Unlike later Mauser models, the top of the receiver ring is unmarked.

Table 1-6 Serbian Serial Number Practice	
1 – 10,000	A1 – A10,000
10,001 – 20,000	B1 – B10,000
20,001 – 30,000	V1 – V10,000
30,001 – 40,000	G1 – G10,000
40,001 – 50,000	D1 – D10,000
50,001 – 60,000	Dj1 – Dj10,000
60,001 – 70,000	E1 – E10,000
70,001 – 80,000	Ž1 – Ž10,000
80,001 – 90,000	Z1 – Z10,000
90,001 – 100,000	I1 – I10,000
100,001 – 110,000	J1 – J10,000

Bayonets were marked on ricasso (refer to Figure 1-2):

GEBR. WEYERSBERG/Solingen

According to the original agreement, the Serbian commission appointed to inspect and receive the arms then left for Oberndorf, see Figures 1-5 and 1-6. By decree, on 1 May 1886 the Serbian government set the price of the infantry rifle at 89 dinars ($17.80).

NOTE: The Commission for inspecting and receiving the arms consisted of the following officers: Pavle Shafaric, Ilija Colak-Antich, Pavle Najdanovich, Pavle Jurisich-Sturm, Kosta-Koka Milovanovich, Yovan Brdarski, Danilo Barkovich, Kosta Kostich, Zivko Kasidolac, Lieutenant (later Captain) Lazar Petrovich, Yovan Pavlovich, Mihailo Rasich as well as the following makers: Vasiljevich, Babich, Parovich, Magazinovich, Milutin Markovich and Milisav Petrovich.

Serbian and Yugoslav Mausers

Figure 1-5. The former Mauser Brothers & Co., Oberndorf, 1886.

The Model 1880 Rifle Deliveries

At the end of 1881, the first shipment of arms was received but the rifles were not immediately issued. The army decided that unit leaders needed to be trained in their use and on 17 December 1882 the first 15-day training course was begun at facilities at Nish and Belgrade. While the

Figure 1-6. Serbian Commission for inspecting the M1880 rifles at the Mauser factory in Oberndorf, left to right: (upper row) Milan (family name unknown) controller at the Kragujevac factory; Lieutenant Danilo Barkovich, Captain Pavle Jurisich-Sturm, Captain Kosta Najdanovich, Zivko Vasiljevich. Middle row: Captain Pavle Shafaric, Major Kosta-Koka Milovanovich. Bottom row: Captain Lazar Petrovich, Captain Kosta Kostich, Captain Yovan Pavlovich.

courses were in progress, the last rifle from the tenth "I" series was completed on 12 February 1884 and received at Belgrade on 27 February. Its job done, the Commission returned to Belgrade on 13 March.

German sources indicate that Serbia received between 100,000 and 120,000 Model 1880 "Kokinka" rifles. Officially, the army had ordered

Serbian and Yugoslav Mausers

only 100,000 Model 1880 rifles, 1,000 spare barrels, and 125,294 other spare parts. But, in 1884 the Gebrueder Mauser & Company announced in the Serbian *Military Gazette* that they "had hundreds more Mauser-Koka rifles for unrestricted sale in their stores at *Sreten Velickovich*, 'The Big Market,'" in Belgrade.

It can be safely stated, then, that the Mauser factory manufactured more than 100,000 completed rifles, refer to Figure 1-1, and intended to use the surplus rifles for commercial sales. The rear sight in these commercial rifles was graduated to 1,000 steps (729.4 yards or 667 meters). Some of the sporting rifles were reserved as presentation pieces and they were elaborately decorated at the Military Technical Institute at Kragujevac, see Figure 1-7. The first presentation rifle, manufactured in 1883, went to the Minister of War, General Milojko Lesjanin. It featured a silver inlaid inscription that read, in Serbian, "General Lesjanin/ Manufactured at The Military Technical Institute, Kragujevac" on the barrel. The rifle is currently in the collection of the Serbian Military Museum in Belgrade.

The remainder of the presentation rifles characterized by their gold and silver inlay, carved stocks, and "AI" (for King Alexander I Obrenovich) the royal cipher and "V.T." (for Military Technical Institute) Kragujevac on the barrel, were used as rewards for winning official shooting competitions. One of these rifles, which was the first prize awarded on the

Figure 1-7. A presentation certificate for a "Kokinka" rifle shooting prize.

Serbian and Yugoslav Mausers

occasion of the seventh public shooting competition in 1894 at Nish, is also in the Serbian Military Museum in Belgrade, see Figure 1-8.

Another example, a Model 1880 10.15 mm, which was the fourth prize at the same competition, was sold by the Swiss auction company Hofman & Reinhart Waffen AG, Zuerich, in 1984 for 3,000 Swiss francs ($1,800).

Figure 1-8. The inscription on the "Kokinka" shooting prize.

Model 1880, Military Use

At the time Serbia acquired the Model 1880 10.15 mm Mauser it was considered to be one of the best rifles in the world, but its service history was not as successful. It was used in combat for the first time in 1885 during the Serbian-Bulgarian War. Believing in the weapon's long-range capability, Serbian soldiers tended to open fire from a distance of 1,000 or more steps (667 meters or 729.4 yards) with poor results. Because of lack of training and ammunition, Serbian troops were not well trained with the Model 1880 rifle, see Figures 1-9 and 1-10. Technological advances involving both smokeless powder and repeating rifles rendered the "Kokinka" obsolete within 10 years of its introduction to the Serbian army.

Figure 1-9. Serbian army privates in service dress. They are armed with the Model 1880 10.15 mm rifle. Photograph dated 1886.

Some Model 1880 rifles were modernized during the first years of the twentieth century; most, however, were relegated to use by third-line draftees or remained in the military stores as surplus.

Serbian and Yugoslav Mausers

After World War I, the few remaining Model 1880 10.15 mm single-shot infantry rifles were converted to fire the Werndl Model 77 (11 x 58 mm R) ammunition. These rifles were stamped on the right side of the rear stock "ΠP/11mm" (the "P" in Cyrillic is "R" in English and denotes a converted weapon). The stamp indicated conversion to calibre 11 mm. Other markings on the weapon occurred during production at Oberndorf.

NOTE: Alexander I Obrenovich (1876-1903), was King of Serbia from 1889 to 1903.

Cavalry and Artillery Carbines Mauser Model 1884

Figure 1-10. Serbian army private in field dress with marching kit. Photograph dated 1886.

In spite of the initial financial problems involving the contract for the Serbian Model 1880 10.15 mm "Kokinka" the Mauser company managed to recover financially and so began to experiment with repeating arms. Peter Paul Mauser succeeded his brother Wilhelm Mauser as managing director and his overriding interest was in the development of repeating arms.

He based his work on the Serbian rifle Model 1880 and equipped it with a tubular 10-shot magazine. His first model (DR Patent No.15,202, dated 16 March 1881) required further improvements. In 1882, some 2,000 modified experimental rifles, designated Type C/82, were sent for field-testing to the German Army garrisons of Darmstadt, Koenigsberg, and Spandau.

As the Serbian commissioners charged with inspecting and accepting the Model 1880 rifles were still at the Mauser factory during this period, Peter Paul Mauser concluded that they represented an opportunity to sell the new repeating rifle concept. Since the previous contract did not include carbines, Peter Paul Mauser had the factory quickly develop a short repeating rifle in 10.15 mm calibre. On 3 November 1883, he sent a sample to Belgrade. The new repeating carbine weapon so impressed the Artillery Committee that on 5 July 1884, Serbia entered into nego-

Serbian and Yugoslav Mausers

tiations with Mauser for the new weapon. Within a month—2 August 1884—Serbia and Peter Paul Mauser concluded an agreement for 4,000 artillery and 4,000 cavalry carbines, designated the Model 1884 10.15 Carbine, see Figure 1-11.

By this time the commission that had inspected and accepted the Model 1880 "kokinka" rifles had returned to Belgrade. The government created a new commission, consisting of experts employed by the Kragujevac factory, to oversee the production of the Model 1884 carbine and they traveled to Germany on 17 October 1884. Production went quickly and a year later, 28 October 1885, the last of the 8,000 carbines were delivered to Serbia.

Identifying Features, the Model 1884 Mauser Cavalry and Artillery Carbines

Figure 1-12 illustrates the main identifying features of the Model 1884 10.15 mm cavalry and artillery carbines.

The pivoting cartridge carrier and the tubular magazine, which ran through a tunnel in the stock, were based on the design used in the Kropatschek Model 1874 Carbine issued to the gendarmerie of Hungary and Bosnia. The cavalry carbine held seven cartridges—five in the magazine, one in the carrier, and one in the chamber. The artillery carbine held eight cartridges—six in the magazine, one in the carrier, and one in the chamber.

The carbine's rear sight had a fixed aperture in the form of a "V" notch that was set to 300 steps (218.8 yards or 200.1 meters). It also had a fixed leaf, hinged at the rear that was graduated from 500 (364.7 yards or 333.5 meters) to 1,400 steps (1,021.2 yards or 933.8 meters). The rear sight leaf had a "V"-notch aperture at the top set for 1,600 steps (1,167.1 yards or 1,067.2 meters).

The cavalry carbine was stocked to the muzzle and the front sight was protected by raised ears on the nose cap, refer to Figures 1-11 and 1-12. The

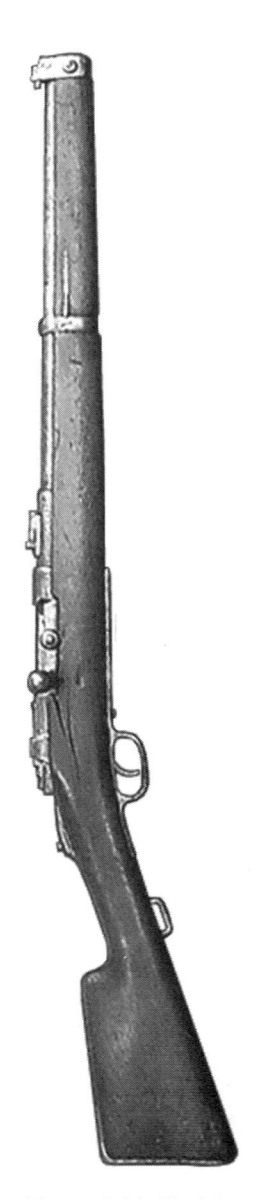

Figure 1-11. Model 1884 10.15 mm Cavalry Carbine.

Serbian and Yugoslav Mausers

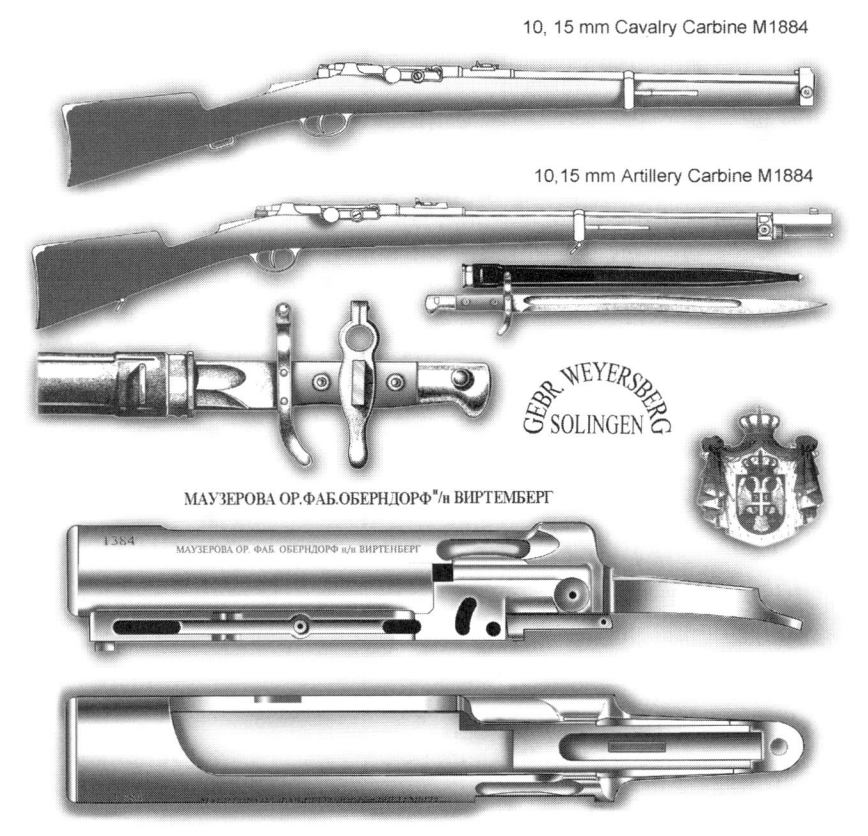

Figure 1-12. Model 1884 10.15 mm Cavalry carbine (above); Model 1884 10.15 mm Artillery carbine and bayonet (below).

cavalry carbine's barrel was 443.6 mm (17.46 inches) long. The stock was held by two barrel bands. The rear barrel band was held in position against the stock step by a band spring. The forward barrel band was held by a cross-bolt that passed through the stock between the barrel and magazine tube. The front band was nearly flush with the muzzle. The trigger guard plate and bow were similar to those used on the rifle. A rectangular staple with rounded ends set into the rear of the trigger guard plate served as the attaching point for the cavalry sling worn by mounted troops. The cavalry carbine did not use a bayonet as cavalry troopers were equipped with swords.

Serbian and Yugoslav Mausers

The artillery carbine's barrel was 534 mm (21.0 inches) or 90 mm (3.5 inches) longer than the cavalry carbine barrel. The barrel protruded past the end of the stock so it could be used to mount the rifle's sword bayonet. The bayonet stud was located on the upper right side of the nose band in a manner similar to the rifle. This upper, or nose, band was identical to that used on the German Model 71/84.

The stocks were made of European walnut and had a high comb and long wrist which help in identifying the two carbines as Serbian Model 1884s. The stock was given an oil finish. The metal furniture received the same deep rust blue finish as the Model 1880 infantry rifle.

The magazine tube was mounted under the barrel and ran through a tunnel in the stock's forearm. A spring was compressed as cartridges were loaded through the open breech into the magazine tube. The spring pushed a follower forward which in turn pushed each cartridge out into the pivoting carrier mounted under the bolt. As the action was cycled, a lug on the bottom of the bolt moved each cartridge off the carrier and into the breech, depressing the carrier at the same time to allow the next cartridge to move into the carrier.

Table 1-7 summarizes the specifications of the Model 1884 cavalry and artillery carbines.

\multicolumn{4}{c}{Table 1-7 Technical Data Model 1884 10.15 mm Carbine for Cavalry and Artillery}			
Model	\multicolumn{3}{l}{Serbian Model 1884 Cavalry and Artillery Carbines}		
Calibre, mm	\multicolumn{3}{l}{10.15 x 63 mm R}		
Rifling	\multicolumn{3}{l}{wedge, 4 grooves, 0.25 mm (0.0098 inch) deep and from 4.6 to 4.0 mm (0-18 to 0.15 inch) wide, breech to muzzle; one turn in 550 mm (21.6 inches), right hand (pitch of 3° 19' 5")}		
Magazine	\multicolumn{3}{l}{under-barrel tube, 5-round capacity (Cavalry) or six (Artillery) plus one cartridge on the elevator and one in the chamber}		
Loading system	\multicolumn{3}{l}{manual insertion of cartridges through floor of bolt-way}		
Dimensions		Cavalry	Artillery
Length overall	mm	959	1,049.4
	inch	37.75	41.3
Barrel length	mm	443.6	534
	inch	17.46	21
Weight	gm	3682	3,880
	lbs	8.1	8.55

Serbian and Yugoslav Mausers

	Table 1-7, cont. Technical Data Model 1884 10.15 mm Carbine for Cavalry and Artillery	
Model	Serbian Model 1884 Cavalry and Artillery Carbines	
Sights	steps (yards/ meters)	(front) unprotected barleycorn; (rear) fixed aperture battle sight for 300 steps (218.8 yards or 200.1 meters) and a large leaf with a sliding extension for 500 to 1600 steps (364.7 to 1,167.1 yards or 333.5 to 1,067.2 meters)
Action		Small-ring Mauser action, 1.32 in. (33.5 mm) diameter receiver ring, 10.43 inches (265 mm) in length, with screw spacing of 9.53 inches (242 mm). Bolt body length 6.53 inches (166 mm). Magazine length 3.23 inches (82 mm). Cavalry carbine not equipped for a bayonet.

Model 1884 Carbines, Markings

Model 1884 10.15 mm carbines were marked with the manufacturer's name and address in Cyrillic on the left side of the receiver, refer to Figure 1-12:

MAUZEROVA OR. FAB. OBERNDORF n/N VIRTENBERG
(Mauser Arms Factory, Oberndorf a/N, Wuerttemberg).

The serial number was marked in Arabic numerals on the left side of the receiver. Unlike other Serbian serial numbers, it did not have a prefix. The serial numbers for the artillery and cavalry models ran sequentially from 1 through 4,000. The receiver ring was unmarked on the top and did not show the Serbian crest or model number, see Figure 1-13. Table 1-8 summarizes all markings on the Model 1884 carbines.

The knife bayonet was marked "GEBR. WEYERSBERG" in a half circle over "SOLINGEN" on the ricasso, refer to Figure 1-12.

Model 1884 10.15 mm Carbines, Problems

Ordnance experts did not take long to discover the repeating carbines' deficiencies, most of which related to the tubular magazine. A description of the carbines published in the *Military Gazette* No. 27, dated 12 July 1887, noted some of the problems. The carbines were slow to load and the cartridges became dangerously hot during prolonged firing due to the magazine's proximity to the barrel. The carbines' center of gravity was too far forward and as the magazine was emptied, shifted to the rear,

Serbian and Yugoslav Mausers

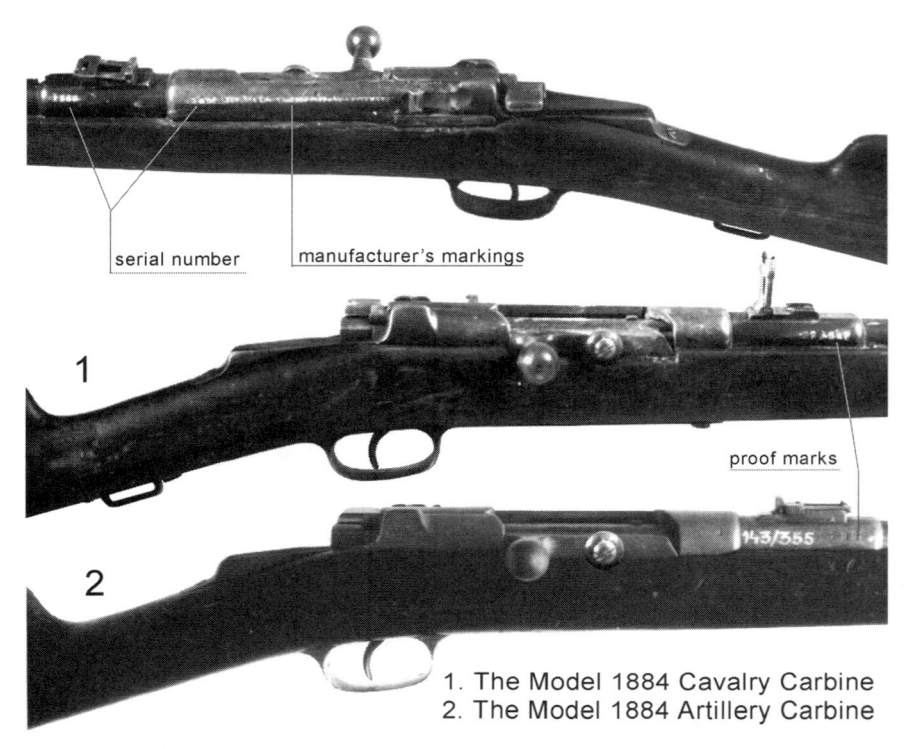

Figure 1-13. Model 1884 Cavalry and Artillery carbines showing the location of the serial number, manufacturer's markings and proof marks.

1. The Model 1884 Cavalry Carbine
2. The Model 1884 Artillery Carbine

interfering with the soldier's attempts to aim and fire rapidly. The recoil acting against the cartridges in the tubular magazine had a tendency to push the bullets into the cartridge case. At the same time, the recoil also caused the cartridges to constrict the magazine spring as far as it would go and when it expanded, shoved the cartridges forward, bullet nose against cartridge base.

The Model 1884 carbines used the Model 1880 10.15 mm centerfire cartridge manufactured with the ogival, or rounded, nose and a sharp jolt against the primer ahead could and did ignite the cartridge ahead in the magazine. German field-testing uncovered these faults and led to the introduction of a new flat-nose bullet in the centerfire cartridge. But this did not happen until after the Serbian contract was completed.

Serbian and Yugoslav Mausers

Markings	Coat of Arms and Model Marking	Manufacturer's Markings	Country Name	Year of Production	Ruler's Monogram	Full Serial Number	Last Three or Four Digits of Serial Number	Proof Mark	Inspection Mark	Acceptance Mark	Calibre
Table 1-8 Markings — Cavalry and Artillery Carbines, Mauser Model 1884											
Barrel:							x	x	x	x	x
Receiver Ring, Top											
Left						x					
Right								x	x	x	
Receiver, Body											
Left	x										
Right								x	x	x	
Bolt Assembly							x		x		
Stock						x					
Trigger Assembly							x		x		
Trigger Guard Plate							x				

Interestingly enough, the problem of pointed bullets in tubular magazines was well known at the time. The Winchester Repeating Arms Company had specified flat-nose bullets for the centerfire cartridges used in their lever action repeating rifle and carbine which also used tubular magazines as early as 1873.

Serbian and Yugoslav Mausers

As a consequence of these deficiencies and other problems, the Serbian government relegated the carbines to noncombatant units such as the gendarmerie. Transfer began on 11 April 1889.

It is believed that only 126 cavalry and 815 Model 1884 10.15 mm artillery carbines remained in service when the First World War began in late August 1914. A number of the Model 1884 carbines marked "11 mm Model 84C" (the "C" in Cyrillic is "S" in English and denoted a Serbian weapon) survived World War I and wound up in the inventory of the new Yugoslav Army. Beginning in 1937, these arms were converted to fire the Gras Model (18)74 cartridge in 11 mm calibre. The carbines so converted were marked "IIP/11mm" (the "P" in Cyrillic is "R" in English and denotes a converted weapon) on the right side of the buttstock. The original manufacturer markings in Cyrillic remained on the left side of the receiver:

MAUZEROVA OR. FAB. OBERNDORF n/N VIRTENBERG
(Mauser Arms Factory, Oberndorf a/N, Wuerttemberg)

The serial number range for the carbines ran from serial numbers 1 through 4,000 without a preceding letter designation. Every major part of the carbine carried the full or partial serial number as was shown in Table 1-8. Mismatched numbered parts are probably the result of repairs or refurbishment.

Financing the Model 1884 Mauser Carbine
By decree, on 1 May 1886 the Serbian government set the price of the cavalry carbine at 78 dinars ($15.60) and the artillery version at 87 dinars ($17.40). The artillery bayonet was the same price as the Model 1880 infantry bayonet.

AMMUNITION FOR THE MODEL 1880/1884, THE 10.15 x 63 MM R
Figure 1-14 provides an illustration of the Serbian 10.15 x 63 mm rimmed, compressed black powder cartridge with a paper-patched bullet, and its military head stamps.

When it first acquired the Model 1880 infantry rifle, the Principality of Serbia did not have its own ammunition-manufacturing facilities. So it ordered the required 10.15 x 63 mm R(imfire) ammunition from the Fabrique Nationale d'Armes de Guerre, Herstal-lèz-Liège (FN). By doing so, they hoped to attract the attention of Belgian investors in the expansion of the Serbian railway.

Serbian and Yugoslav Mausers

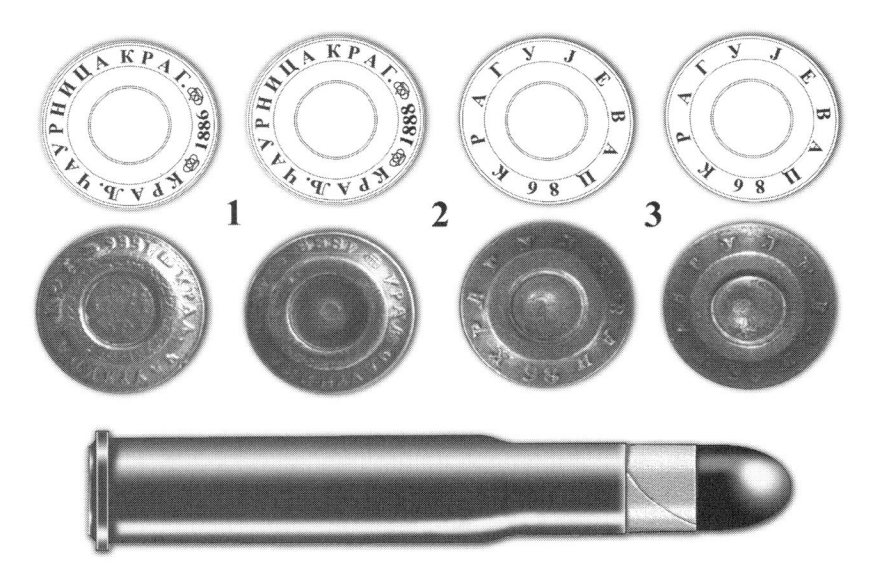

Figure 1-14. The 10.15 x 63 mm R cartridge. Headstamps: 1, Kraljevska Caurnica Kragujevac (Royal Cartridge Case Workshop, Kragujevac), 1884-1890; 2 and 3, Kragujevac, 1890-1914.

A problem arose when it was realized that the testing of the new rifle conducted at Oberndorf, Germany, between 1881 to 1884 was accomplished using German-made ammunition. These cartridge cases were 63.5 mm (2.5 inches) long and were produced by Lorenz in Karlsruhe, Germany. They bore the head stamp "L" or "D." Powder for the cartridges was manufactured at Pulverfabrik Rottweil factory at Rottweil on Neckar with bullets cast at the Mauser factory. During the trials, it was found that a more suitable length for the cartridge case was 63 mm (2.48 inches). Based on the results of German development, the Military Technical Institute at Kragujevac began to purchase the necessary machinery and dies to produce cartridges for the Model 1880 10.15 mm rifle domestically. The machines were installed in 1884 in a new building which became part of the Cartridge Case Workshop (Caurnica) complex and was soon in operation.

But differences in performance between domestic ammunition and the imported ammunition soon became apparent. The Belgian cartridge used a charge of brown prismatic powder (braunem prismatischen Pulver) SGP (Serbisches-Gewehr-Pulver) from Pulverfabrik Duenerberg at Besenhorst, that achieved an initial muzzle velocity of 512 mps (1,680

Serbian and Yugoslav Mausers

fps). Domestic ammunition used a powder made at Stragari, Serbia, which produced a muzzle velocity of only 460 mps (1,509 fps) and had an average range of 3,250 meters (3,554 yards) with a penetrating depth of 200 mm (7.87 inches) at this extreme range.

In addition to these difficulties, ammunition production at Kragujevac still had not reached the projected target of 100,000 to 120,000 cartridges per day after two years. Consequently, the Serbian government felt compelled to order 5 million 10.15 x 63 mm R cartridges from the then-small factory, K.k. Hof-Metallwaren-und Patronenfabriken von Seraphin Keller in Hirtenberg, Germany.

NOTE: As with the Mauser company, this contract made it possible for Seraphim Keller and Sons to invest money in factory development. Over the years, the company evolved into the famous Hirtenberger Patronen-Zuendhuetchen-und-Metallwarenfabrik Aktiengesellschaft, vormals Keller & Co., which is still in operation today.

Burst Cartridge Cases

Another problem that developed was, in fact, quite common to all military ammunition made during the 1870s and 1880s—burst cartridge cases. The technology for drawing brass into cylinders able to withstand breech pressures of 12,000 psi or more, was only then being developed. As a consequence, extracting damaged and stuck cartridge cases from the rifle was of great concern.

An article in *Military Gazette*, No. 32, 10 August 1884, titled, "The Regulation on Extraction of the Caps from Burst Cartridge Cases Including Their Cleaning," described the extraction of burst Model 1880 10.15 mm ammunition cases from the breech using a hydraulic device designed by Mata Blaskovich, a lathe operator in the Military Technical Institute at Kragujevac, see Figure 1-15.

His detailed drawings were attached to these regulations and proved very useful. The National Assembly realized the obvious savings in maintenance and repair costs and decided on 30 March 1888 to reward the talented lathe operator with 800 dinars ($160, no small sum in those days). His company was so proud of Blaskovich's invention that it was included in its display at the 1889 World Exhibition in Paris.

Nevertheless, Blaskovich's device was not entirely new. In July 1876, the Bavarian Army War Office issued "Regulations on Handling a Hydraulic Device for Extracting Caps from Burst Cartridge Cases Model 1871" for the German Model 71 Mauser. This regulation was also published in the Bavarian *Military Gazette*, No. 3. The illustrations that accompanied the regulations were drawn by Colonel Hans-

Serbian and Yugoslav Mausers

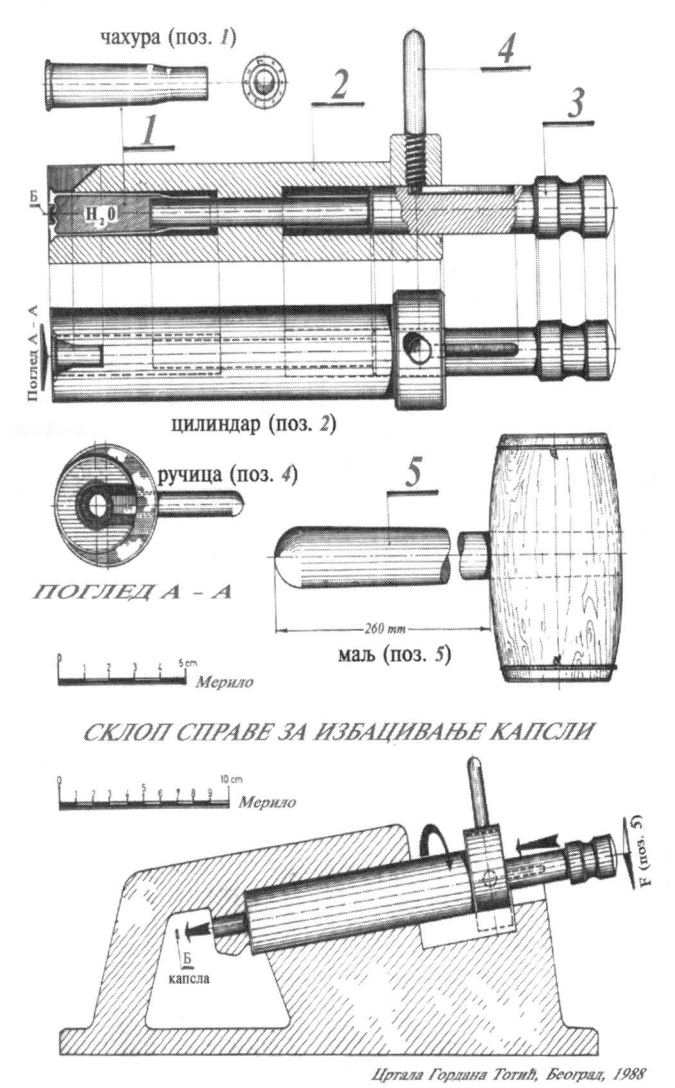

Figure 1-15. The Blaskovich device for extracting burst cartridges.

Jurgen Stoll and were identical to those in the later Serbian Regulation A/TNo 762. Obviously Blaskovich had seen either Stoll's drawings or the Bavarian device itself and only adjusted the device to the Serbian-produced cartridges.

CHAPTER 2
MAUSER RIFLES MODEL 1899, MODEL 1899/07, MODEL 1910, AND CARBINE MODEL 1908 AND DJURICH-MAUSER-KOKA MODEL 1880/07

By the start of the 1890s, two potential enemies of Serbia, Austria and Turkey, had adopted small-bore repeating rifles using smokeless powder ammunition. The Artillery Committee of the Serbian Army recognized that their forces needed better weapons. In 1891, the War Office established a commission to choose an appropriate modern repeating rifle system. A two-year detailed inspection of various makes followed and led to the recommendation to adopt the Spanish 7 mm Mauser Model 1893 (Mauser Espanol/Modelo 1893), produced by Loewe in Berlin. But because of Serbia's poor financial position, she was unable to secure the loans necessary to purchase the new rifles.

MODEL 1899 MAUSER RIFLE

For most of the decade, the Serbian government sought ways to finance the massive purchase of the 110,000 magazine rifles. The purchase price was estimated at 6,730,000 dinars ($1,346,000) for the rifles, 152,000,000 dinars ($30,400,000) for ammunition, and 1,000,000 dinars ($200,000) for ammunition pouches and accessories. Not until 1899 did the Union Bank in Vienna, Austria, agree to fund a contingency contract with the German firm, Deutsche Waffen-und Munitionsfabriken A.-G, to deliver 90,000 of the required 150,000 new rifles. See Appendix B for a complete discussion of financial negotiations.

Ordnance Officer Nedeljko Vuckovich was appointed president of a commission to inspect and accept the new rifles by the Serbian Minister of War. By June 1899, the factory delivered and the Serbian commission accepted the first one hundred of the new Model 1899 7 x 57 mm Mauser rifles and 20,000 7 x 57 mm cartridges.

But the shipment from Germany to Serbia had to go by railroad through Austria, and the Austro-Hungarian government held up the shipment for two months while arguments over transit permits were settled.

Not until March 1900 did Serbia receive 22,000 of the new Model 1899 rifles. By June, an additional 39,000 rifles were received in Serbia, which allowed for the issue of 420 rifles per regiment. By the end of the year, DWM had delivered the full 90,000 rifles.

Serbian and Yugoslav Mausers

Identifying Features of the Model 1899

The Serbian Model 1899 7 x 57 mm Mauser design was similar to the Spanish Model 1893 Mauser and the Chilean Model 1895 Mauser rifles, see Figure 2-1. Figure 2-2 provides a contemporary disassembled view of the Spanish Model 1893 Mauser.

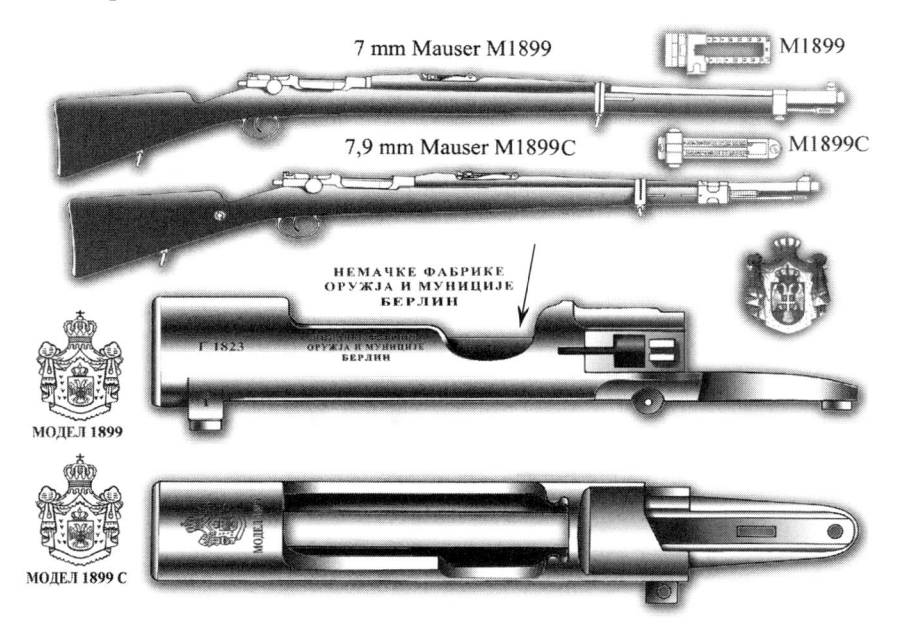

Figure 2-1. Above, Model 1899 Mauser 7 mm Infantry rifle; below, converted to fire the 7.92 x 57 mm Mauser cartridge and redesignated the Model 1899C.

The Serbian Mauser differed from both the Spanish and Chilean models in that the left side of the receiver rail had a thumbcut to make it easier to load ammunition in clips (arrow in Figure 2-1).

Other features of the Model 1899 Serbian Mauser Rifle were: 1) One-piece rifle stock without a nose cap and with a straight wrist. The stock nose was enclosed by the forward barrel band. Unlike its predecessor, the Model 1899 was fitted with a handguard extending from the receiver ring forward to the rear barrel band. The handguard had a rectangular hole for the rear sight. 2) The Mauser-style bayonet mount was located on the bottom of the barrel band. 3) The rifle was built on the so-called "small-ring" receiver and used a Mauser turnbolt action with two lugs (see Figure 2-3); 4) the bolt handle protruded to the right at a ninety-degree angle to the receiver. 5) The bolt body had two lugs which

Serbian and Yugoslav Mausers

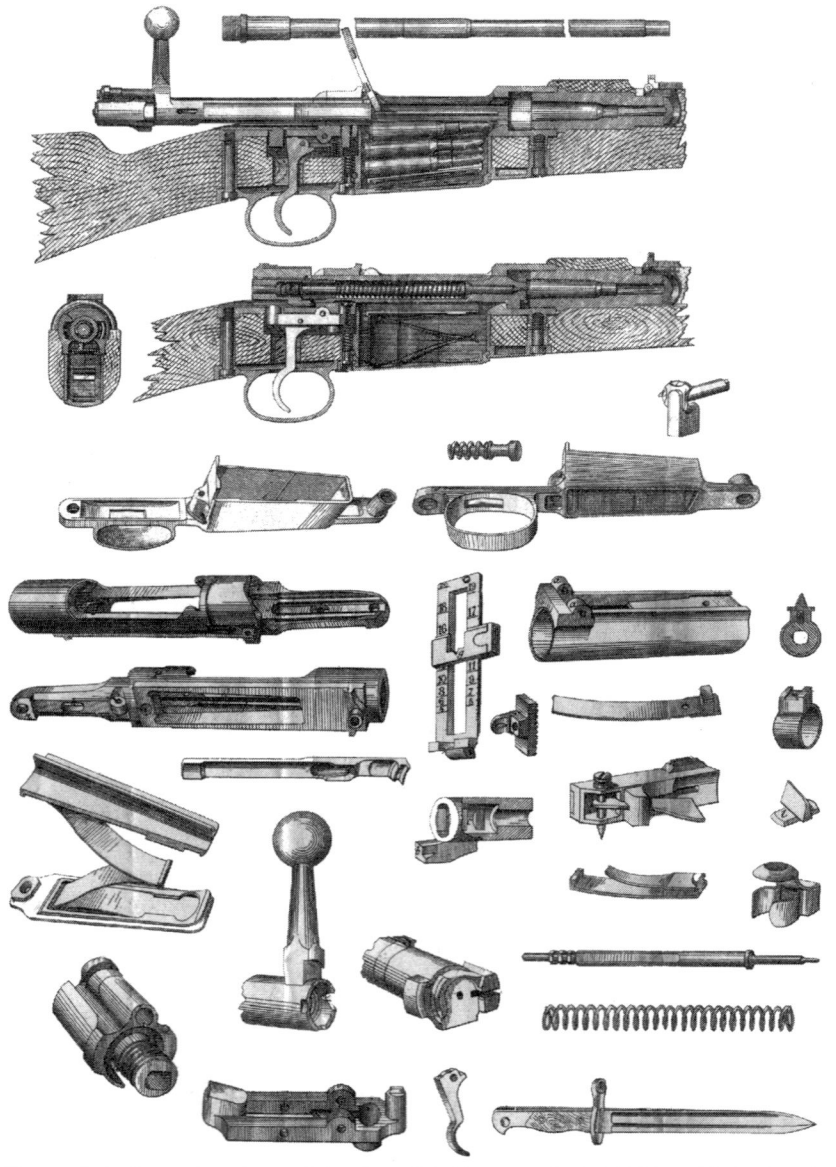

Figure 2-2. Parts drawings of the Spanish Mauser Model 1893 published in the annual, *Ratnik*. The Model 1893 was one proposed design for the new Serbian infantry rifle.

Serbian and Yugoslav Mausers

Figure 2-3. Above, the Model 1899 bolt assembly. Below, the Model 1899/1907 bolt assembly.

turned into matching cuts in the receiver wall to lock the bolt into battery. 6) The left lug was split to allow it to pass over the ejector stud. 7) The forward sling swivel was mounted on the bottom of the rear barrel band and the rear sling swivel on the bottom of the stock. 8) The rear sight leaf was mounted on the barrel forward of the receiver ring. It had a leaf hinged at the rear. The leaf was graduated to 2,000 meters (2,187 yards). The aperture was a "V"-notch on the moveable elevation slide or sight bar. 9) The box magazine was enclosed entirely within the stock and held five cartridges. 10) The calibre of the rifle was 7 x 57 mm.

The Model 1899 was equipped with a knife-bayonet 402 mm (15.8 inches) long overall, see Figure 2-4. The Serbian bayonet was based on a double-edged design with a Mauser-style hilt and was reminiscent of the later Mexican Model 1910 bayonet. The scabbard was steel.

After the rifles had been delivered and had seen a few months' service, it became clear that some modifications were needed. Headspacing of the cartridge was found to be excessive in many rifles. The cartridge chambered on its shoulder but the chamber proved to be too long, allowing a gap of as much as 3 mm (0.118 inch) between it and the bolt face (see Figure 2-5). This excessive gap sometimes caused the cartridge to split, releasing gas into the receiver and causing it to burst. Moreover, the extractor proved to be too long and was prone to breakage. And Serbian unit commanders reported that the catch on the elevation slide was easily damaged.

Engineers from the Military Technical Institute immediately set to work to eliminate these defects. Philip Petrovich designed a new sight slide. Other gunsmiths determined that if a groove, or circular ring for

Serbian and Yugoslav Mausers

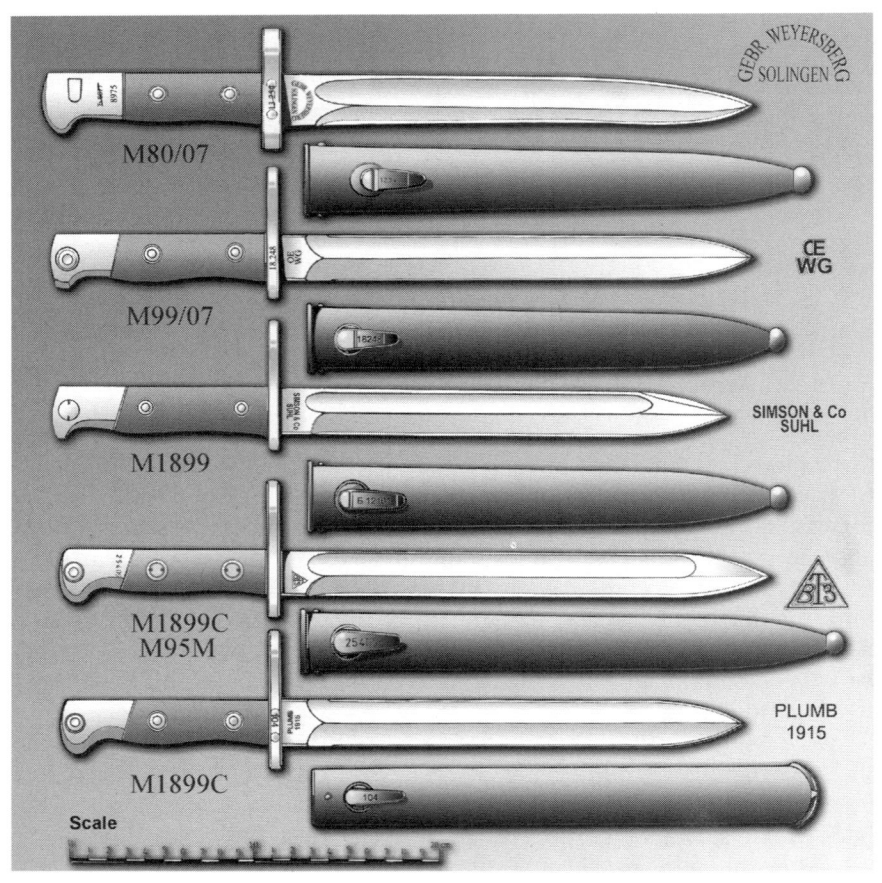

Figure 2-4. Bayonets for the Serbian Mausers: From the top, Model 1880/07, Model 1899/07, Model 1899, Model 1899C and Model 1895M, and the Model 1899C which was manufactured by Fayette R. Plumb, Co. of Philadelphia, PA, USA.

the cartridge base was machined into the rear of the chamber, they could correct the excessive headspace. The ring provided clearance for the extractor head and allowed the cartridge to fully enter the chamber, thus eliminating the 3 mm (0.118-inch) gap. This change in the small-ring Model 1899 Mauser action assured that the chambered cartridge base was fully supported. Modifying the rifles cost 2.43 dinars ($0.48) each.

Receivers with the machined groove or ring at the breech to fully enclose the cartridge case head were referred to as "safety breech" models. Peter Paul Mauser adopted the Serbian innovation and first manufactured

Serbian and Yugoslav Mausers

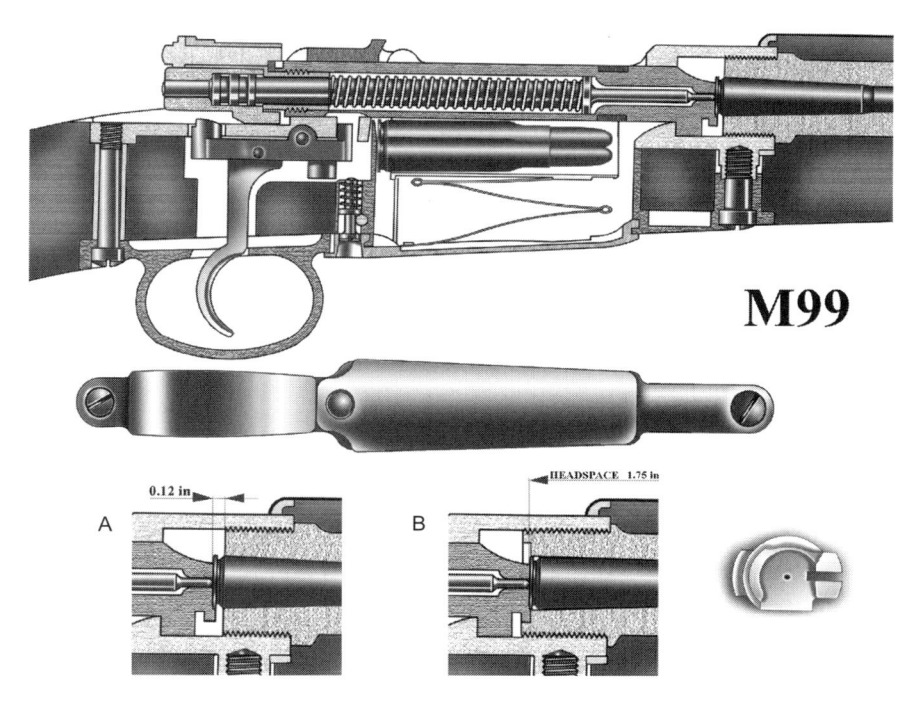

M99

Figure 2-5. Cross section drawing of the Model 1899 Mauser 7 x 57 mm Infantry rifle. Insets A and B below show how the excessive headspace problem was solved by machining a ring in the breech to fully enclose the cartridge base and bolt face—the "Ring of Steel."

this type of action in 1910 with the Serbian (Model 1910) and Mexican (Model 1912) contract Mauser rifles. The standard German Model 1898 left approximately 2.5 mm (0.1 inch) of the cartridge head exposed and unsupported. The system also came to be known as the "Ring of Steel." Technical data is summarized in Table 2-1 below.

Model 1899 Rifle Markings
Refer to Figure 2-1 and see Figure 2-6 for the location and style of markings. The most reliable way to identify the Serbian Model 1899 7 x 57 mm rifle is by the Serbian crest marked on the receiver, above the legend, "Model 1899" in Cyrillic. The left side of the receiver was marked in Cyrillic,

NEMACKE FABRIKE/ORUZJA I MUNICIJE/BERLIN

Serbian and Yugoslav Mausers

(German Arms and Ammunition Factory)

The serial number was stamped on the left side of the receiver rail, ahead of the manufacturer's name in the form of the Cyrillic letter designating the number 10,000 and the number of the rifle within that production run, i.e., B 9326, signifies the 29,326th Model 1899 rifle. Refer to Table 1-6 for the Serbian serial-numbering practices.

THE MODEL 1899/07 RIFLE AND MODEL 1908 CARBINE

The entire contract with DWM for the Serbian Model 1899 7 x 57 mm Mauser rifle was sufficient to outfit little more than one half of the Serbian army's wartime strength. An inventory dated 12 April 1904 for the Chief of the General Staff indicated that the first mobilization group would require 108,136 rifles, 162,204,000 rounds of ammunition, and 321,193 ammunition pouches. But the inventory also revealed that the army had only 90,019 Model 1899 rifles, 85,991,186 cartridges and a third of the required number of pouches.

Without the domestic capacity to produce sufficient modern rifles, the Serbian government had to initiate another round of arms contracts to purchase more foreign-made weapons. In early September 1903, Isidor Goldstein, an agent for Bannerman's of New York City, offered to provide 10,000 original Spanish Mauser Model 1893 rifles along with 3,500,000 rounds of 7 x 57 mm calibre ammunition at 30 dinars ($6) per rifle and 40 dinars ($8) per 1,000 rounds of ammunition for a total of 440,000 dinars ($88,000).

The rifles had been part of the arms seized by the United States Army after defeating Spanish forces in Cuba and the Philippines in 1898. The rifles had been purchased from the U.S. Government by the well-known

Figure 2-6. Model 1899 Mauser 7 x 57 mm rifle receiver markings.

Serbian and Yugoslav Mausers

New York military goods dealer, Francis Bannerman. He also purchased the entire supply of Spanish Model 1895 carbines which had also been seized as well.

In the summer of 1902, Bannerman contracted with the National Armory at Springfield, Massachusetts, to fully recondition these weapons for sale in Europe during the summer of 1903. He immediately fixed on Serbia as a potential customer because that nation had already begun arming its military with almost identical rifles.

In pushing the deal, Bannerman sent several specimens of the Model 1893 rifle to his agent, Isidor Goldstein, who was then headquartered in Vienna. Austria and Serbia were at loggerheads over various trade disputes at the time; the Austrian *K.u.K. Evidenzbuero* in Vienna seized the sample shipment and warned the Austrian military attaché in Belgrade, Major Joseph Pomiankowski, about the imminent transaction. Major Pomiankowski immediately notified the Austro-Hungarian Army's Chief of Staff, General Friedrich von Beck. The Austrian government protested the sale and negotiations with Bannerman were suspended.

Again, the Serbian government was forced to seek a loan from the international bankers to complete their weapons program. After several years of negotiations they were finally able to sign a contract with the Oesterreichische Waffenfabrik-Gesellschaft, Steyr, or OEWG, a major Austrian weapons manufacturer (Figure 2-7) for the manufacture and purchase of 30,000 rifles and 10,000 carbines. See Appendix B for a complete discussion of the financial arrangements.

The new rifles and carbines were chambered for the 7 x 57 mm cartridge, and the contract included a large quantity of 7 mm rifle barrels for use as spares and to rebarrel other existing rifles. The new infantry rifles received the Serbian designation Model 1899/07 and the carbines, the designation Model 1908. Table 2-1 provides the technical data that describes the Models of 1899, 1899/07 and the Model 1908 Carbine.

Figure 2-7. Section No.VIII of the Arms Factory, Steyr, Austria, at the end of the 19th century.

Serbian and Yugoslav Mausers

\multicolumn{4}{c}{Table 2-1 Technical Data, Models of 1899, 1899/07 and Carbine Model 1908}			
Model	Serbian Mauser Infantry Rifle M1899	Serbian Mauser Infantry Rifle M1899/07	Serbian Mauser Carbine M1908
Calibre	7 x 57 mm		
Rifling	concentric, 4 grooves, 0.125 mm (0.0049 inch) deep and 4.1 mm wide (0.16 inch); 1 turn in 220 mm, right hand (pitch 5° 42' 30")		
Bolt	2 locking lugs		
Magazine	internal staggered-column box, capacity 5 rounds		
Loading system	charger, or single rounds		
Length overall	1,142.2 mm 44.97 inches	1,142.2 mm 44.97 inches	944 mm 37.16 inches
Barrel length	738 mm 29.05 inches	738 mm 29.05 inches	448 mm 17.64 inches
Weight	4000 gm 8.82 lbs	4000 gm 8.82 lbs	3180 gm 7.01 lbs
Sights	(front) unprotected barleycorn; (rear) a leaf with a sliding extension for 300 to 2,000 m (328 to 2,187 yds)	(front) protected barleycorn; (rear) a leaf with a sliding extension for 300 to 1,500 m (328 to 1,640.4 yds)	(front) unprotected barleycorn; (rear) a leaf with a sliding extension for 300 to 2,000 m (328 to 2,187 yds)

Identifying Features, Model 1899/07 Rifle and Model 1908 Carbine
The Model 1899/07 7 x 57 mm infantry rifle was virtually identical to the Model 1899 7 x 57 mm rifle. It is chiefly distinguished by its different markings, see Figures 2-8, 2-9, and 2-10. Marked in Cyrillic on the receiver,

AUSTRIJSKA/ ORUZNA FABRIKA /STAJER
(Austrian Arms Factory, Steyr)

Serbian and Yugoslav Mausers

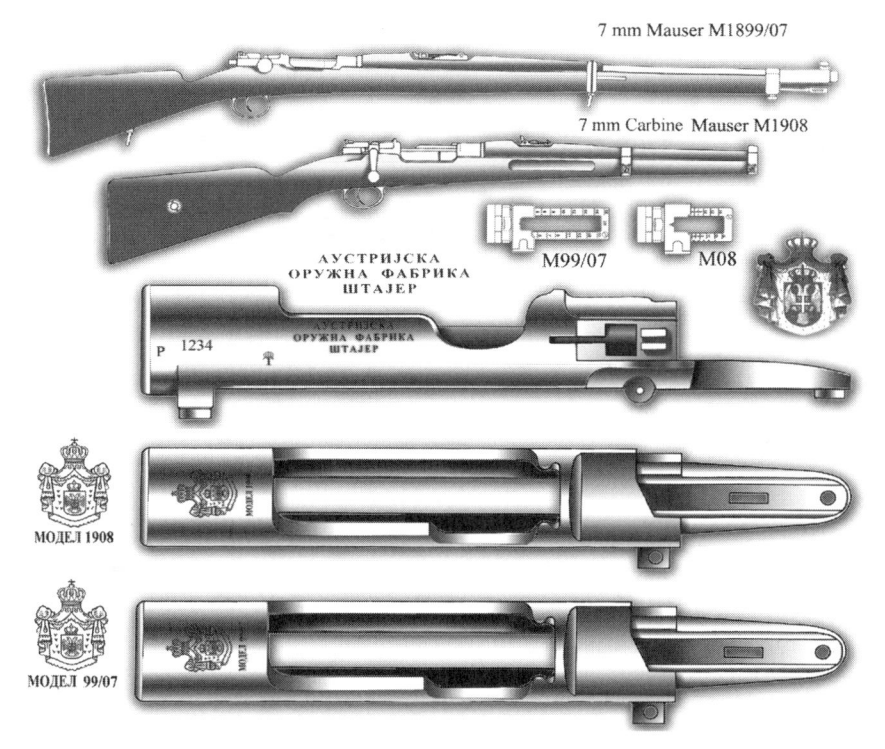

Figure 2-8. Above: Model 1899/07 Mauser 7 x 57 mm Infantry rifle. Below: Model 1908 Mauser 7 x 57 mm Cavalry carbine.

The Model 1899/07, as manufactured by OEWG, incorporated the changes made to the Model 1899 to eliminate the headspace and rear sight problems. It was also equipped to accept the Model 1899 bayonet, refer to Figure 2-4. Table 2-2 describes the finishes applied to each part of the Model 1899/07 rifle.

Figure 2-9. Model 1899/07 Mauser receiver markings.

51

Serbian and Yugoslav Mausers

Figure 2-10. Receiver marking on the receiver ring, Model 1908 7 x 57 mm Cavalry carbine. Note the distinguishing sight groove in the handguard ferrule.

The Model 1908 7 x 57 mm carbine was a shortened version of the rifle but with a full-length stock that reached to the muzzle and finger grooves on either side, refer to Figure 2-8 and see Figures 2-11 and 2-12. The Model 1908 Carbine was the only Serbian carbine to have a pistol-grip stock. The forward barrel band was secured by a transverse screw through the stock nose. The front sight was protected by ears on the forward barrel band similar to those used on the Spanish Model 1892 and 1895 carbines. The carbine rear sight was graduated to 1,600 meters (1,749.8 yards). The carbine was not equipped to accept a bayonet. Both rifle and carbine had five-shot magazines completely enclosed within the stock.

Table 2-2 describes the finishes applied to each part of the Model 1908 Carbine.

Table 2-2 Finishes Models of 1899, 1899/07 Rifles and Model 1908 Carbine	
Parts	Finish
All wood parts	Sanded smooth and oiled
Barrel	Blued, hot salt bath method after polishing and degreasing
Receiver	Blued, hot salt bath method after polishing and degreasing
Front Band, Lower Band	Blued, hot salt bath method after polishing and degreasing
Metal plate on the underside of the butt	Blued, hot salt bath method after polishing and degreasing
Swivels	Blued
Butt Plate	Finished in-the-white
Front Sight Blade	Blued by heating

Serbian and Yugoslav Mausers

Figure 2-11. Model 1908 Mauser 7 x 57 mm Cavalry carbine.

Table 2-2, cont.
Finishes
Models of 1899, 1899/07 Rifles and Model 1908 Carbine

Parts	Finish
Rear Sight Base	Blued, hot salt bath method after polishing and degreasing
Sight Leaf	Blued
Slide	Blued
Slide Catch	Blued
Bolt Parts	Finished in-the-white
Bolt Stop	Blued by heating
Trigger Guard and Floor Plate	Blued, hot salt bath method after polishing and degreasing
Trigger	Blued
Sear	Finished in-the-white
Follower	Blued
Front and Rear Guard Screws	Blued
Magazine Spring	Blued spring steel
Floor Plate Catch	Blued by heating

THE SERBIAN MODEL 1910 MAUSER RIFLE

The purchase of the Austrian-made Model 1899/07 7 x 57 mm rifles and the Model 1908 7 x 57 mm carbines satisfied only a third of the additional small arms needed by the Serbian military in the event of war. As a stopgap measure, the Serbian government directed that surviving single-shot 10.15 mm Mauser-Milovanovich Model 1880 ("Koka") rifles be converted to fire the 7 x 57 mm cartridge from a five-shot magazine (described below). When the converted rifles fell far short of expectations, the Government decided on another foreign purchase.

Serbian and Yugoslav Mausers

Figure 2-12. A detachment of the 4th Cavalry Regiment during the 1912 campaign against Turkey. The troopers are equipped with the Model 1908 carbine.

Serbian and Yugoslav Mausers

In 1909, the National Assembly approved another French loan for 150,000,000 dinars ($30,000,000) of which 95,000,000 dinars ($19,000,000) were for firearms.

During this time, there were notable developments in the country's economic and foreign policy situation. Serbia and Austro-Hungary had signed a new agreement on 22 July 1910 to re-establish commercial relations. But just as it seemed that the way was now clear to purchase the remainder of the 110,000 rifles needed from the Austrian company, Steyr, the government in Vienna suddenly restricted meat imports from Serbia.

In protest, the Serbian government granted the contract instead to the German concern, Deutsche Waffenfabrik-und Munitionsfabriken A.-G (DWM). The contacts with DWM were made by Georg Brankovich, an agent of the Serbian government through the companies, Norddeutsche Lloyd, and Friedrich Krupp AG. Brankovich conferred through Georg Luger with Max Kosegarten and Paul von Gontard, directors of DWM's plants in Berlin, as well as Felix Haenisch, Director of the Martinkenfelde, Germany plant. Serbia's order totaled 32,000 Model 1910 Mauser rifles based on the design of the now famous German Model Gewehr (18)98.

Identifying Features, Model 1910 Mauser
The Serbian Model 1910 7 x 57 mm rifle looked, at first glance, very much like the Model 1899 7 x 57 mm long rifle, but with a tangent rear sight and the large-ring Mauser 1898 action. It featured "the ring of steel" full cartridge base support used in the Model 1899 and Model 1899/07 rifles, and the Model 1908 carbine. The bolt had three locking lugs.

The Model 1910 breech fully enclosed the cartridge base while the standard German Model 1898 left about 2.5 mm (0.1 inch) of the cartridge base exposed and unsupported. All subsequent Yugoslavian Mauser models were constructed with the same system to fully enclose the cartridge.

The Model 1910 rifle was equipped with the standard Mauser Model 1898 five-round magazine assembly, refer to Figure 2-11. Whereas the stocks on previous models (except the Model 1908 carbine) had a straight wrist, the Model 1910 rifle had a pistol grip. But, unlike the Mauser Model 1898, it did not have a recoil cross-bolt in the stock. This was judged unnecessary with the milder-recoiling 7 x 57 mm Mauser cartridge. The Model 1910 was equipped with a modern tangent rear sight and the bolt, like that of the Model 1898 German design, had three locking lugs. The rifle was very popular with the troops.

Serbian and Yugoslav Mausers

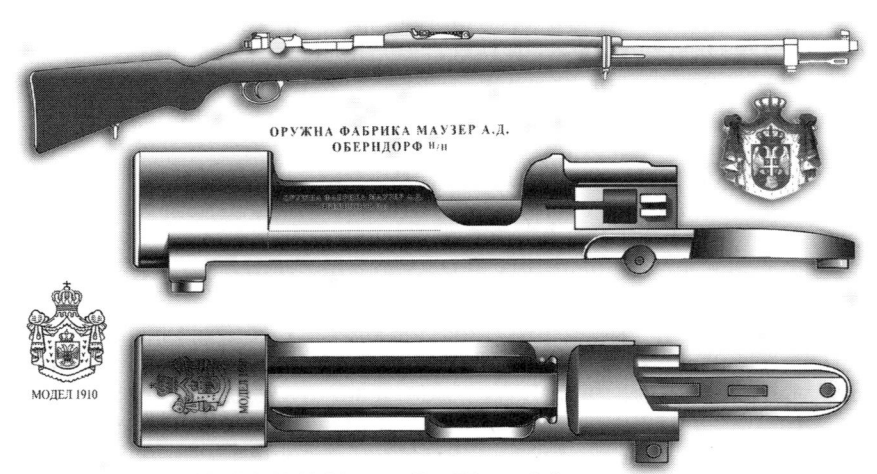

Figure 2-13. The Model 1910 Mauser 7 x 57 mm Infantry rifle.

Table 2-3 includes the specifications for the Model 1910 Mauser Rifle and Table 2-4 describes the finish applied to each part.

Table 2-3 Technical Data, Model 1910 Mauser Rifle	
Calibre	7 x 57 mm
Rifling	concentric, 4 grooves, 0.125 mm (0.0049 inch) deep and 4.1 mm (0.16 inch) wide; 1 turn in 220 mm, right hand (pitch 5° 42' 30")
Bolt	3 locking lugs
Magazine	internal staggered-column box, capacity 5 rounds
Loading system	charger, or single rounds
Length overall	1,142.2 mm 44.97 inches
Barrel length	738 mm 29.05 inches
Weight	4,000 gm 8.82 lbs
Sights	(front) unprotected barleycorn; (rear) a leaf with a sliding extension for 300 to 2000 m (328 to 2,187 yds)

Serbian and Yugoslav Mausers

Table 2-4
Markings
The Mauser Rifle M1910 7 mm

	Also see Table C-1, C-2, C-3, C-4, C-5, C-6, C-7, C-8	Coat of Arms and Pattern Mark	Manufacturer's Markings	Country's Name	Year of Production	Ruler's Monogram	Full Serial Number	Last Three or Four Digits of Serial Number	Smokeless Powder Proof Mark	Inspection Mark	Acceptance Mark	Calibre
Barrel:							x		x	x	x	x
Receiver Ring:												
Top		x										
Left							x		x			
Right										x		
Receiver												
Left			x									
Right										x	x	
Bolt Assembly												
Bolt Body												
Bolt Handle Base										x		
Bolt Handle Neck							x					
Bolt Handle Ball									x			
Firing Pin												
Cocking Piece								x		x		

Serbian and Yugoslav Mausers

Table 2-4, cont.
Markings
The Mauser Rifle M1910 7 mm

Also see Table C-1, C-2, C-3, C-4, C-5, C-6, C-7, C-8	Coat of Arms and Pattern Mark	Manufacturer's Markings	Country's Name	Year of Production	Ruler's Monogram	Full Serial Number	Last Three or Four Digits of Serial Number	Smokeless Powder Proof Mark	Inspection Mark	Acceptance Mark	Calibre
Bolt Sleeve with Gas Shield							X		X		
Bolt Sleeve Stop											
Safety Lever							X				
Extractor											
Extractor Collar											
Ejector											
Bolt Stop									X		
Trigger Assembly											
Trigger									X		
Sear-fork									X		
Trigger Guard Plate						X					
Magazine Floor Plate							X		X		
Stock											
Butt Plate						X					
Front Band, Lower Band									X		

Serbian and Yugoslav Mausers

Table 2-5 Finishes, Model 1910 Mauser Rifle	
Part	**Finish**
All wood parts	Sanded smooth and oiled
Barrel	Blued, hot salt bath method after polishing and degreasing
Receiver	Blued, hot salt bath method after polishing and degreasing
Front Band, Lower Band	Blued, hot salt bath method after polishing and degreasing
Metal plate on the underside of the butt	Blued, hot salt bath method after polishing and degreasing
Swivels	Blued
Butt Plate	Finished in-the-white
Front Sight Blade	Blued by heating
Rear Sight Base	Blued, hot salt bath method after polishing and degreasing
Sight Leaf	Blued
Slide	Blued
Slide Catch	Blued
Bolt Parts	Finished in-the-white
Bolt Stop	Blued by heating
Trigger Guard and Floor Plate	Blued, hot salt bath method after polishing and degreasing
Trigger	Blued
Sear	Finished in-the-white
Follower	Blued
Front and Rear Guard Screws	Blued
Magazine Spring	Blued spring steel
Floor Plate Catch	Blued by heating

Serbian and Yugoslav Mausers

Model 1910 Mauser Rifle Markings

The Model 1910 7 x 57 mm Mauser was marked with the Serbian royal crest on the receiver ring above the legend, in Cyrillic, "Model 1910." The left side of the receiver rail was marked in Cyrillic:

ORUZNA FABRIKA MAUZER .AD.OBERNDORF n/N
(Mauser Arms Factory, Oberndorf am Neckar)

Refer to Figure 2-11 and see Figure 2-14 below for the location of the Cyrillic markings.

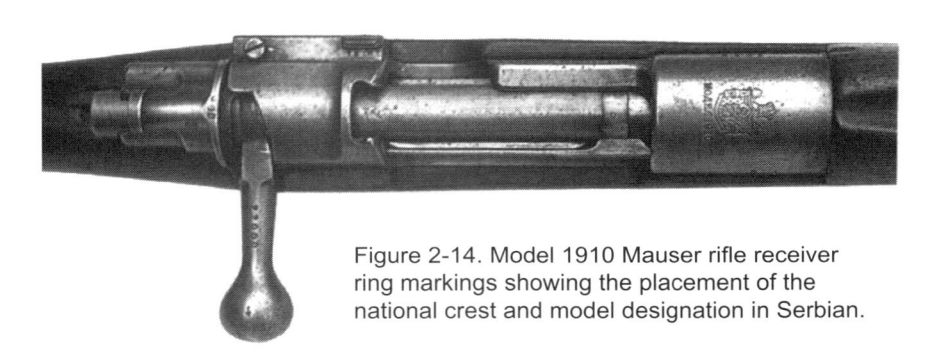

Figure 2-14. Model 1910 Mauser rifle receiver ring markings showing the placement of the national crest and model designation in Serbian.

The Oberndorf factory began deliveries of the Model 1910 rifles that year. By December, 4,800 rifles had reached Serbia with the remainder delivered by mid-1911.

Serbian records indicate that it now had 156,000 repeating infantry rifles and 10,800 carbines of the Mauser system in 7 x 57 mm calibre in inventory. Added to this were the 43,000 converted Model 1880 "Kokinka" pattern rifles for a total of 199,000 infantry rifles in 7 x 57 mm calibre.

Georg Luger, the Chief-Constructeur b.d. Deutsche Waffen- und Munitionsfabriken A.-G, Dipl Ingen and designer of the famous Luger pistol, confirmed this number in a confidential memo sent to a Montenegrin officer, Ilija Hajdukovich, dated 22 May 1912, wherein he stated that "Serbia bought almost 180,000 rifles (Mausers) in 7 mm calibre but many were lost mainly during the *Komite-Chetnik's* fighting in Macedonia," see Figure 2-15.

Serbian and Yugoslav Mausers

NOTE: *Komitagy* or *Komite* were the members of special paramilitary formations trained to conduct operations in the enemy's rear areas. The name derives from the French word *comité* (committee), in other words, from the "Committee for Actions in Macedonia." The term *Chetnik* (pl. *Chetniks*) is also used as a synonym; it derives from the Turkish term *çete* denoting the guerrilla units.

THE MODEL 1899, 1899/07, AND 1910 AFTER WORLD WAR I

When the new Kingdom of Yugoslavia standardized the arms carried by its constituent states after 1929, it retained the Mauser system and modified the Model 1899, Model 1899/07, and Model 1910 rifles to use the 7.92 x 57 mm calibre cartridge, the standard German military cartridge from 1898 to 1945, instead of the Serbian 7 x 57 mm cartridge. These conversions ultimately included shortening the rifles to carbine length, the adoption of the later Model 1924 sights, and the replacement of the barrel bands and sling swivels.

Figure 2-15. *Komita* Dragisha Stojadinovich is equipped with the Serbian Model 1899 infantry rifle in this photograph taken at Belgrade in 1906. Komitagy were paramilitary forces who operated behind enemy lines.

In the nomenclature of the Yugoslav Army, the rifles were listed as:

7.92 mm Carbine Model 99C
7.92 mm Carbine Model 99/07C
7.92 mm Carbine Model 1910C

The suffix Cyrillic "C" (English "S") denoted a weapon converted from Serbian arsenal inventories, see Figure 2-16. Many of the Model

Serbian and Yugoslav Mausers

Figure 2-16. Receiver and bolt assembly of the Model 1899C Mauser rifle. The Serbian letter "C" (S in English) signifies that the rifle was rebuilt to fire the 7.92 x 57 mm Mauser cartridge.

Figure 2-17. Receiver ring marking of the Model 1899C Mauser rifle.

1910s in 7 x 57 mm calibre that had remained unconverted were later seized by the Germans in April 1941 and redesignated as:

7 mm Gew.221(j)

Weapons converted to 7.92 x 57 mm by the Kingdom of Yugoslavia had the manufacturer's name removed from the left side of the receiver and added below the Serbian coat of arms on the receiver ring (see Figure 2-17) with suffix "C" (S) denoting the weapon's Serbian origins:

Model 1899C
Model 99/07C
Model 1910C

SUMMARY, SERBIAN MODEL 1899, 1899/07 AND 1910 MARKINGS
All Serbian Mauser rifles and carbines were marked with the Serbian coat of arms consisting of a two-headed eagle surmounted with a crown against the backdrop of a baldachin, see Figure 2-18. The coat of arms was stamped on the receiver ring above the model marking in Cyrillic which will read: "Model 1899," "Model 99/07," "Model 1910," and for the carbine, "Model 1908." On the left side of the receiver will be found the manufacturer's name in Serbian:

Model 1899
  NEMACKE FABRIKE/ORUZJA I MUNICIJE/BERLIN

Serbian and Yugoslav Mausers

(German Arms and Ammunition Factory)

Model 99/07 and the Model 1908
AUSTRIJSKA/ORUZNA FABRIKA/STAJER
(Austrian Arms Factory, Steyr)

Model 1910
ORUZNA FABRIKA MAUZER.AD. OBERNDORF n/N
(Mauser Arms Factory, Oberndorf on Neckar)

Figure 2-18. The royal coat of arms served as the national crest for Serbia until 1918.

All three models remained in service well into the 1930s, Figure 2-19.

POSTWAR YUGOSLAVIAN MODEL DESIGNATIONS AND MODEL MARKINGS

When the newly created Kingdom of Yugoslavia standardized the arms carried by its constituent states in the 1930s, it retained the Mauser system and modified the Model 1899, Model 1899/07, and Model 1910 rifles to the 7.92 x 57 mm calibre cartridge, refer to Figures 2-1, 2-8 and 2-12. These conversions ultimately included shortening rifles to carbine length, the adoption of the Model 1924 sights, and the replacement of the barrel bands and sling swivels.

In the nomenclature of the Yugoslav Army, the rifles were listed as the "7.92 mm carbines, Model 99C," "Model 99/07C," and "Model 1910C." The suffix Cyrillic "C" (English "S") denoted a weapon converted from Serbian arsenal inventories. Many of the 7 mm Model 1910s were, in turn, seized by the Germans in April 1941 and redesignated as the "7 mm Gew.221(j)."

Weapons converted to 7.92 mm by the Kingdom of Yugoslavia had the manufacturer's name removed from the left side of the receiver. The model designation was added below the Serbian coat of arms on

Serbian and Yugoslav Mausers

Figure 2-19. Members of the Royal Guard equipped with the M1908 carbine and M1899 infantry rifle. The bayonets were made by the American company, Fayette R. Plumb, Co. of Philadelphia and St. Louis.

Serbian and Yugoslav Mausers

the receiver ring (refer to Figure 2-17) with suffix "C" (S) denoting the weapon's Serbian origins:

> Model 1899C
> Model 90/07C
> Model 1910C

The Djurich-Mauser-Koka Model 1880/07

By the latter decades of the 1800s, Serbian ordnance officers were well aware of the advantages in firepower to be gained from repeating rifles, even as they were selecting a single-shot breech-loading rifle with which to equip their army. It was felt that repeating-firearms technology was too new and uncertain—a feeling that was reflected in numerous other armies, including that of the United States. Only Serbian Army Major Kosta Milovanovich disagreed. He had, in the course of working with Wilhelm Mauser on the initial arms purchase, become familiar with the German achievements in the field of converted box magazines. When Mauser patented detachable magazines in the period 1879-1880, particularly DR Patent 41,375, Major Milovanovich resolved to convert Serbia's single-shot Model 1880 rifles to magazine-fed, repeating rifles.

Other arms manufacturers such as Krnka, Mannlicher, and Werndl were attempting to do the same. But their efforts were less than successful and they concluded that there was no future in such conversions. But Major Milovanovich, with the design of the successful Model 1880 behind him, was confident that he could succeed where others had failed. By 1882, he had filed patents DRP Nr. 19,673 and Nr. 21,397 in Austria, Italy, France, and Germany for two types of an "adjustable box magazine" applicable to the turnbolt rifle systems of the French Gras, the German Mauser Model 71 and Model 78/80, and the Russian Berdan Model 1870, No. 2, see Figure 2-20. The latter weapon was originally developed by the American inventor and Civil War commander of Berdan's Sharpshooters, Hiram Berdan.

Now General Milovanovich also claimed that, in theory, his conversion increased muzzle velocity and gave the rifles a sustained rate of fire of 30 shots a minute. The increased muzzle velocity claim depended on changing the calibre of a converted single-shot rifle from 10.13 mm to a smaller calibre.

Most Serbian ordnance officers, however, felt that such conversions were of limited value and dismissed the effort. For example, in the "Notes" section of *Military Gazette* No. 48, dated 27 November 1882,

Serbian and Yugoslav Mausers

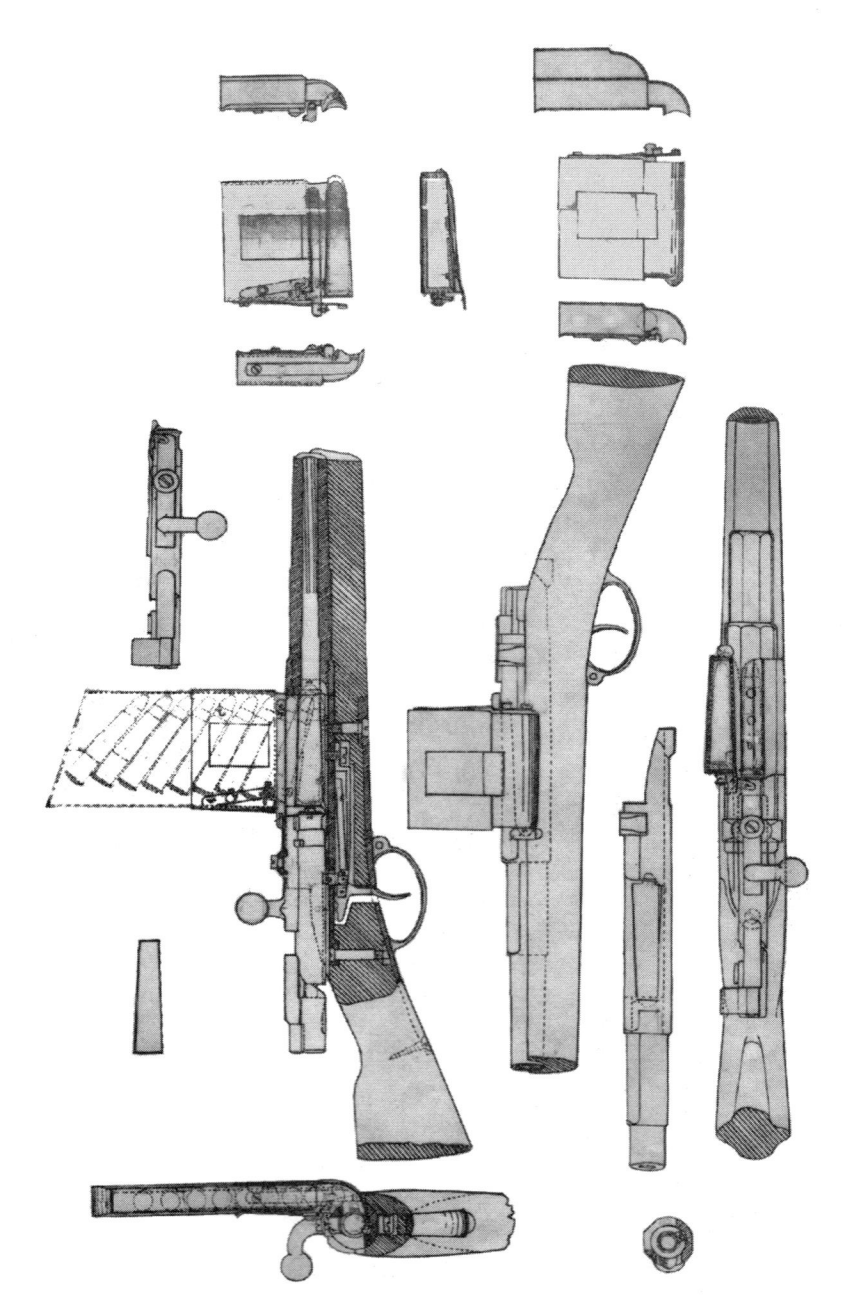

Figure 2-20. Box magazine patent drawing by Kosta Milovanovich as published in the *Military Gazette*, 1884.

Serbian and Yugoslav Mausers

"Koka's" (General Kosta Milovanovich's nickname) inventions received only the briefest of mention. Although rebuffed, Kosta Milovanovich was not discouraged. With renewed enthusiasm and the expenditure of large amounts of money, he devoted himself to developing a repeating rifle of his own invention, Figure 2-21.

In 1891, he submitted his weapon to the Artillery Committee; it was tested with the new Russian 7.62 mm Model 1891 Mosin-Nagant and a number of other modern repeating rifles. After the conclusion of the

Figure 2-21. Patent drawing for the repeating rifle Model 1891 based on the Koka Milovanovich system, published in the journal, *Ratnik*, 1891.

two-year trial, the commission chose the Mauser Model 1893 design that ultimately became the Serbian Model 1899 7 x 57 mm Mauser bolt-action repeating rifle, described earlier.

The failure left the respected General in a difficult financial position. He had no means of paying off the debt of 20,000 golden dinars ($4,000) that he had incurred. But "as a sign of national respect" the Serbian government presented him with this sum on 6 February 1896.

Although General Milovanovich was a visionary and ahead of his time, his concept for converting single-shot rifles into repeating rifles would be realized nearly twelve years later when Serbia's finances made

Serbian and Yugoslav Mausers

it almost impossible to purchase a large quantity of repeating arms abroad.

Vindication
The Army concluded, in a report dated 14 April 1902, that it needed 69,170 of the 10.15 mm Model 1880s for II Class Mobilization inductees. It had on hand 95,354 Model 1880s, which left a surplus of 26,184 rifles.

A later report in April 1904 gave a more accurate count of 93,132 Model 1880s in inventory of which the Army needed only 55,625 rifles in 10.15 mm calibre. Army leaders decided to convert the remaining 37,507 to fire the now-standard 7 x 57 mm smokeless powder ammunition. Offers were solicited from the original manufacturer, Waffenfabrik Mauser AG, as well as the French company Manufacture d'Armes Châtellerault, and the Hungarian arms firm Fegyvergyár in Budapest.

In September 1903, four modified Model 1880 rifles arrived from Budapest and passed a firing trial of 6,000 rounds. Four months later, Mauser submitted three modified rifles for testing. The Serbian government set the cost for each conversion at 22 dinars ($4.40) but not until 1906 did the Assembly finally take up the issue.

Then, after a particularly bitter debate in which many representatives had voted in favor of the Hungarian offer, the government decided to give the project to a domestic manufacturer, the Military Technical Institute, at Kragujevac. On 6 June 1906, the Council of State approved a loan of 6,000,000 dinars ($1,200,000) to purchase the necessary machinery. Events progressed swiftly; a year later, 50,000 7 mm barrels had been purchased from the arms factory at Steyr, Austria, in conjunction with the contract for the Model 99/07 rifle and Model 1908 carbine. Preparations for the conversion at the Military Technical Institute were completed by January 1908.

Figure 2-22. Colonel Gojko Djurich (seated) at the workshop where the Model 1880 Mauser 10.15 mm infantry rifle was converted to the Model 1880/07 to fire the 7 x 57 mm Mauser cartridge.

Serbian and Yugoslav Mausers

Developing the methodology for converting to the new five-shot magazine as well as other details of organization and supervision was the responsibility of the Gunsmith Shop's manager, retired Army Lieutenant Colonel Gojko Djurich, see Figure 2-22. Under his supervision, the Kragujevac factory modified between 100 and 120 rifles a day and, by 12 May 1910, completed 34,000 conversions. By 4 March 1911, a total of 43,000 rifles were finished plus an additional 5,493 rifles before the project concluded. In the end 50,132 unaltered Model 1880s remained in storage. Technical data for the Model 1880/07 conversion are contained in Table 2-5.

\multicolumn{2}{c}{**Table 2-6** **Technical Data,** **Model 1880/07 Mauser Rifle**}	
Caliber	7 x 57 mm
Rifling	concentric, 4 grooves, 0.125 mm (0.0049 inch) deep and 4.1 mm (0.16 inch) wide; 1 turn in 220 mm, right hand (pitch 5° 42' 30")
Bolt	3 locking lugs
Magazine	projecting in-line box, 5-rounds capacity
Loading System	charger, or single rounds
Length Overall	1,027 mm 40.4 inches
Barrel Length	738 mm 29.05 inches
Weight	4,510 gm 9.94 lbs
Sights	(front) unprotected barleycorn; (rear) a leaf with a sliding extension for 300 to 2,000 m (328 to 2,187 yds)

The modification from a 10.15 mm single-shot rifle to a magazine-feeding repeating rifle firing the 7 x 57 mm cartridge required changes to the receiver, bolt, and trigger mechanism, and the installation of a new barrel, see Figures 2-23 and 2-24. The new box magazine held five cartridges and was located beneath the receiver. The magazine protruded below the stock for half of its length. Since the weapon was fitted with

Serbian and Yugoslav Mausers

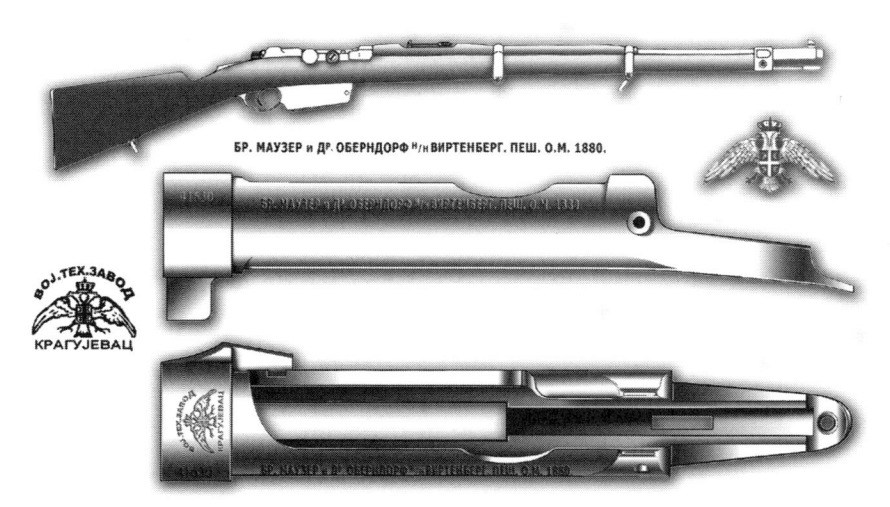

Figure 2-23. The Model 1880 Mauser 10.15 mm rifle converted to the Mauser-Djurich system (Model 1880/07) 7 x 57 mm rifle.

the new 7 mm calibre barrels from Steyr, the rifle had the ballistic features of the Model 1899/07, and a rear sight similar to that used on the Model 1899/07 rifle was installed. See Table 2-7 for a description of the markings and Table 2-8 for the finishes applied to the Model 1880/07 Mauser Rifle.

Table 2-7, Markings The Djurich-Mauser-Koka M1880/07 7 mm											
	See Also Table C-1, C-2, C-3, C-4, C-5, C-6, C-7, C-8	Coat of Arms	Manufacturer's Markings	Country's Name	Year of Production	Ruler's Monogram	Full Serial Number	Last Three or Four Digits of Serial Number	Smokeless Powder Proof Mark	Inspection Mark	Acceptance Mark	Calibre
Barrel								x	x	x	x	x
Receiver Ring											
Top	x	x										
Left						x		x	x			

Serbian and Yugoslav Mausers

Table 2-7, Markings, cont.
The Djurich-Mauser-Koka M1880/07 7 mm

See Also Table C-1, C-2, C-3, C-4, C-5, C-6, C-7, C-8	Coat of Arms	Manufacturer's Markings	Country's Name	Year of Production	Ruler's Monogram	Full Serial Number	Last Three or Four Digits of Serial Number	Smokeless Powder Proof Mark	Inspection Mark	Acceptance Mark	Calibre
Right											
Receiver											
Left		x*									
Right			x*						x	x	
Bolt Assembly											
Bolt Body											
Bolt Head							x, x*				
Bolt Handle Base							x*				
Bolt Handle Neck						x					
Bolt Handle Ball								x			
Firing Pin											
Cocking Piece											
Bolt Sleeve											
Safety Lever							x*				
Extractor									x		
Ejector									x		
Bolt Stop									x		

Serbian and Yugoslav Mausers

Table 2-7, Markings, cont.
The Djurich-Mauser-Koka M1880/07 7 mm

	See Also Table C-1, C-2, C-3, C-4, C-5, C-6, C-7, C-8	Coat of Arms	Manufacturer's Markings	Country's Name	Year of Production	Ruler's Monogram	Full Serial Number	Last Three or Four Digits of Serial Number	Smokeless Powder Proof Mark	Inspection Mark	Acceptance Mark	Calibre
Trigger Assembly												
Trigger										x		
Sear-fork										x		
Trigger Guard Plate								x				
Magazine Floor Plate								x				
Stock												
Butt Plate												
Front/Low-Band									x*			

x* = old markings (M1880)

Table 2-8
Finishes,
M1880/07 Infantry Rifle

Part	Finish
All wood parts	Sanded smooth and oiled
Barrel	Blued, used seven steam cabinet treatments
Receiver	Blued, used a hot salt bath method after polishing and degreasing
Front Band, Lower Band	Blued, used a hot salt bath method after polishing and degreasing

Serbian and Yugoslav Mausers

Table 2-8, cont. Finishes, M1880/07 Infantry Rifle	
Part	Finish
Metal plate on the underside of the butt	Blued, used a hot salt bath method after polishing and degreasing
Swivels	Blued
Butt Plate	Blued, used a hot salt bath method after polishing and degreasing
Front Sight Blade	Blued by heating
Rear Sight Base	Blued, used a hot salt bath method after polishing and degreasing
Sight Leaf	Blued
Slide	Blued
Slide Catch	Blued
Screws	Finished with a black coating
Bolt Parts	Finished in-the-white
Trigger Guard, Magazine Housing and Floor Plate	Blued, used a hot salt bath method after polishing and degreasing
Trigger	Blued
Sear	Finished in-the-white
Follower	Blued
Front and Rear Guard Screws	Blued
Magazine Spring	Blued spring steel

The Models of 1880/07 were the first rifles to bear the markings of the Military Technical Institute of Kragujevac, refer to Figure 2-23 and see Figure 2-25. The Mauser factory's name (transliterated from the Cyrillic) as:

BR. MAUZER I Dr. OBERNDORF n/N VIRTENBERG. PES. O.1880

and remained on the left side of the receiver, but the receiver ring was marked by the two-headed Serbian eagle with outstretched wings (which differed from the eagle on stamped imported Mausers) and the inscription:

Serbian and Yugoslav Mausers

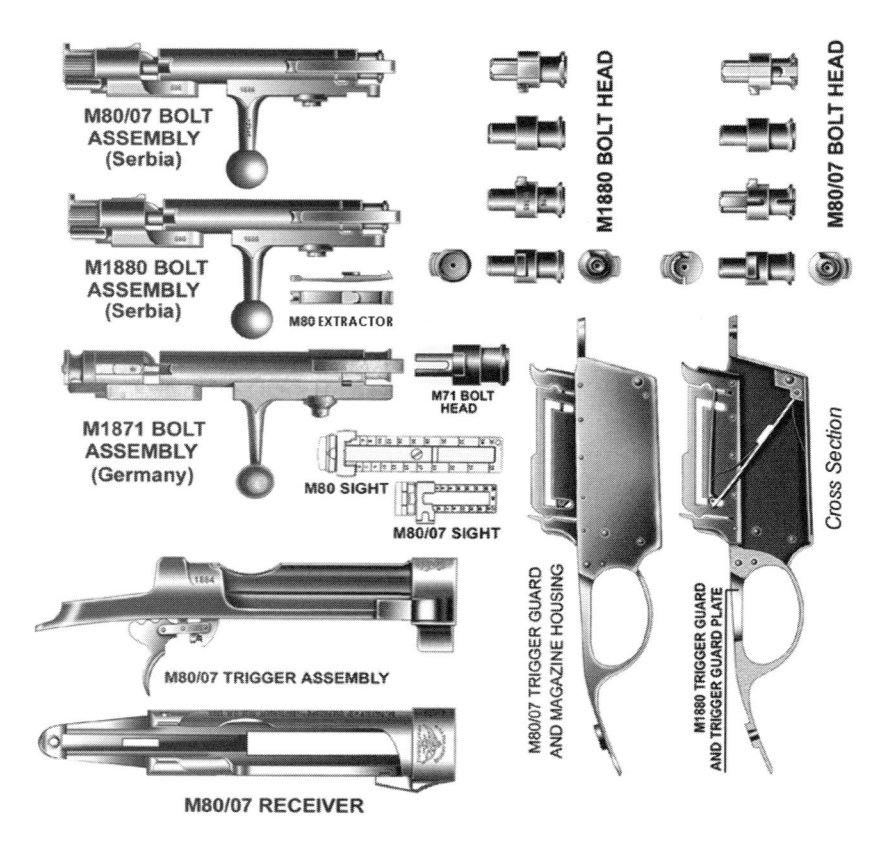

Figure 2-24. The Model 1880 10.15 mm rifle parts affected by conversion to the Model 1880/07 Mauser-Djurich system to fire the 7 mm Mauser cartridge. Also shown is the original German Model 1871 bolt assembly on which the Serbian Model 1880 rifle was based.

VOJ. TEH. ZAVOD/KRAGUJEVAC
(Military Technical Institute/Kragujevac)

Serbian soldiers did not like the converted rifles. For instance, during some of the bitterest operations of the First World War, the Commander of the Serbian First Army, General Petar Boyovich, informed the Supreme Command on 4 October 1914 that the converted weapon "throws too high" and that his soldiers distrusted the converted rifles so much that they were deliberately breaking them. A month later,

Serbian and Yugoslav Mausers

Figure 2-25. Receiver markings on the Model 1880/07 Mauser-Djurich rifle.

on 19 November 1914, the Commander of the Third Army, General Pavle-Yurisich-Shturm, reported that "a large number of the Model 80/07 were functioning badly."

But Serbia's opponent did not share this low opinion and held the Model 80/07 in high regard. In the German journal *Schuss und Waffe* for 1916-1917, one contributor named Wandolleck wrote that it was "a good and useful rifle for combat and was capable of using interchangeable ammunition of other models."

Other Serbian commanders must have agreed, as the Model 1880/07 remained in service until 31 December 1937 when, by decision of the Minister of the Army and Navy, these rifles were given to the Sokol's Association for practice and training with live ammunition. The "Sokol" (Falcon) was a large pre-WWII Serb/Yugoslav youth organization that emphasized physical training and fitness and the development of outdoor survival skills.

7 x 57 MM AMMUNITION

In 1899, DWM transferred its contract for the manufacture of 45,000,000 cartridges in 7 x 57 mm to the Belgian firm, Fabrique Nationale d'Armes de Guerre, Herstal-lèz-Liège, with ordnance Colonel Djordje Markovich as the controller. The Belgian firm delivered this ammunition a year later.

In 1900, the Serbian government, fearful of being dependent on foreign suppliers for its ammunition, decided to begin manufacturing its own small-arms cartridges, see Figure 2-26. By 1902, the Serbian Obilicevo Powder Plant had installed the necessary machinery to produce

Serbian and Yugoslav Mausers

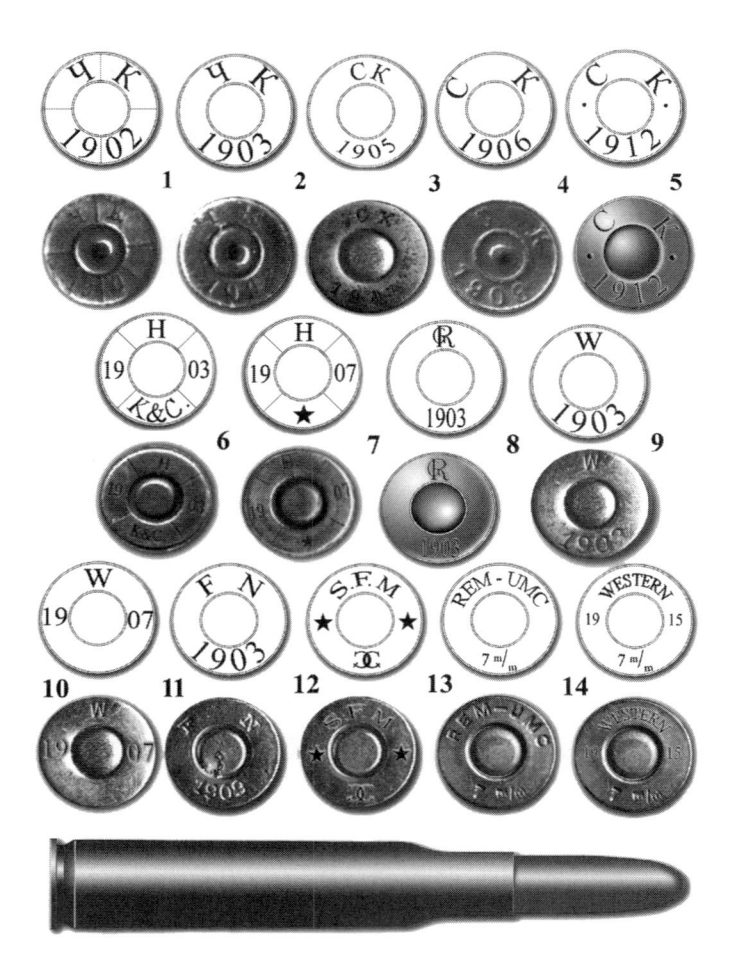

Figure 2-26. Ammunition issued by the Serbian armed forces for the 7 x 57 mm Model 1899 rifle, showing headstamps and cartridge: 1- 2: Caurnica, Kragujevac (Cartridge Case Workshop, Kragujevac) 1899–1903; 3-5: Srbija, Kragujevac (Serbia, Kragujevac), 1903-1914; 6-7: Hirtenberger Patronen-Zuendhuetchen-und-Metallwarenfabrik Aktiengesellschaft, vormals Keller & Company; 8: Georg Roth Fabriken fuer Gewehr und Geschuetz-Patronen, Geschosse, Huelsen, Zuender und Zuendhuetchen Aktiengesellschaft of Vienna; 9-10: Manfred Weiss Aktiengesellschaft, Budapest; 11: Fabrique Nationale d'Armes de Guerre, Herstal-lèz-Liège; 12: Societé Française des Munitions-Gévelot et Gaupillat-Paris; 13: Remington Arms – Union Metallic Cartridge Co., Bridgeport, Connecticut, USA; 14: Western Cartridge Company, East Alton, Illinois, USA.

Serbian and Yugoslav Mausers

Figure 2-27. The Hirtenberg ammunition factory in the second half of the 19th century.

a smokeless powder based on the German firm, Pulverfabrik Rottweil's (Rottweil am Neckar), formula. Raw materials, primarily cellulose, came from K.u.K. Pulverfabrik Blumau at Steinfeld, a division of the Austrian K.u.k. Munitionsfabrik in Woellersdorf.

Once it mastered the production of smokeless powder, the Military Technical Institute began manufacturing cartridges at the rate of 50,000 per day. This was considered this amount to be inadequate for the Army's needs and pressed the government to make another foreign purchase. Serbia subsequently signed a contract for 45,000,000 rounds of ammunition and 100,000,000 ammunition pouches with the firms Hirtenberger Patronen-Zuendhuetchen-und-Metallwarenfabrik Aktiengesellschaft, vormals Keller & Company (see Figures 2-27, 2-28 and 2-29), Georg Roth Fabriken fuer Gewehr-und Geschuetz-Patronen, Geschosse, Huelsen, Zuender und Zuendhuetchen Aktiengesellschaft of Vienna, and Manfred Weiss Aktiengesellschaft, Budapest. Hirtenberg AG's contract was for 670,000 dinars ($134,000) while Weiss AG's contract for 9,000,000 rounds was worth 1,004,400 dinars ($200,880). The largest contract, however, went to Georg Roth AG. The choice of the firms was greatly dependent on the influence of Oesterreichischen Creditanstalt, a limited liability company that guaranteed payment with Serbian state coupons.

Hirtenberg delivered its cartridges on 11 June 1903 and Weiss AG delivered its ammunition a day later.
On the eve of the First World War, Serbia ordered 50,000,000 more 7 x 57 mm cartridges from the Societé Française des Munitions-Gévelot et Gaupillat-Paris, or SFM. In 1916, the British Adriatic Mission con-

Serbian and Yugoslav Mausers

tracted for ammunition with several American firms for the Serbian Government. Remington Arms and Union Metallic Cartridge Company of Bridgeport, Connecticut, and Western Cartridge Company of East Alton, Illinois, together delivered 70,000,000 7 x 57 mm rounds of ammunition and 1,000,000 Model 99 clips.

Figure 2-28. The Serbian Commission for inspecting and receiving 7 x 57 mm Model 1899 ammunition at the Hirtenberg factory.

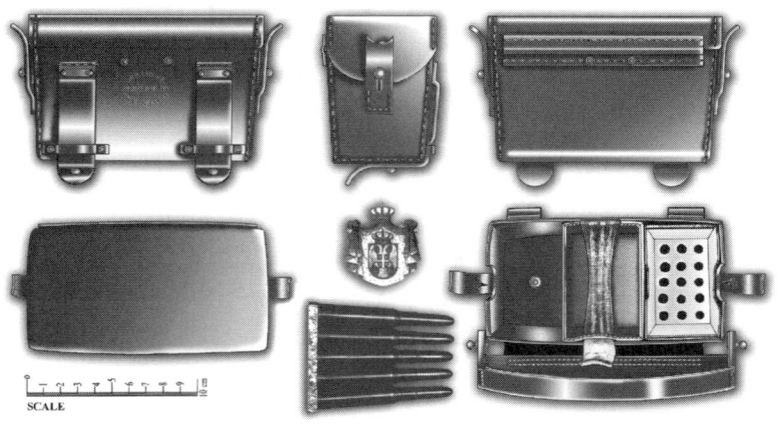

Figure 2-29. Model 1899 rifle ammunition pouch for the 7 x 57 mm cartridge. Fifteen rounds in three clips were carried in each pouch.

CHAPTER 3
TURKISH MAUSERS
IN SERBIAN AND YUGOSLAV SERVICE

TURKISH MODEL 1887 RIFLE AND CARBINE
Turkey began its military modernization program very late in the nineteenth century. It desperately needed new modern small arms and equipment. The government's German advisers informed the Porte, Turkey's ruler, that Germany and Serbia had already decided to adopt repeating firearms with tubular magazines. The Turkish Army's favorable experience with the Winchester Model 1866 repeating rifle during the Russo-Turkish War (1877), combined with the influence of its German advisers, led Istanbul in 1886 to contract with Mauser for the manufacture of 500,000 Model 71/84 infantry rifles with the tubular magazine and 50,000 Model 1887 carbines. Both were chambered to fit the new Turkish 9.5 x 60R mm rimmed cartridge, see Figure 3-1.

TURKISH MODEL 1890 7.65 x 53 MM MAUSER RIFLE
But, while awaiting the Turkish contract, Peter Paul Mauser had patented a number of innovations, which improved on the basic German version of the M71/84 which was designated the Model 1887. He replaced the bolt handle locking mechanism with two opposed locking lugs, mounted a variation of the Lee single-column box magazine in the trigger guard plate to feed cartridges vertically into the breech, and chambered the rifle for a small-bore smokeless powder 7.65 x 53 mm rimless cartridge. He designated it the Model 1889 Mauser rifle, refer to Figure 3-1. The first country to purchase the new rifle, in that year, was Belgium.

Impressed by a demonstration of the new rifle, the Porte changed the 1886 contract to acquire the Model 1889 instead of the improved M71/84. He required that the rifle be chambered for the 7.65 x 53 mm cartridge.

Turkey had already received 220,000 rifles of the Model 71/84 pattern and 4,000 of the carbines under the original contract. The remaining 280,000 infantry rifles and 45,000 carbines were delivered according to the new "Belgian" pattern and were designated the Model 1890 7.65 x 53 mm rifles.

NOTE: The 7.65 x 53 mm rimless cartridge was also adopted by Argentina, Bolivia, Colombia, Ecuador and Peru. It was available in the United States and Canada into the 1930s and is still available in many

Serbian and Yugoslav Mausers

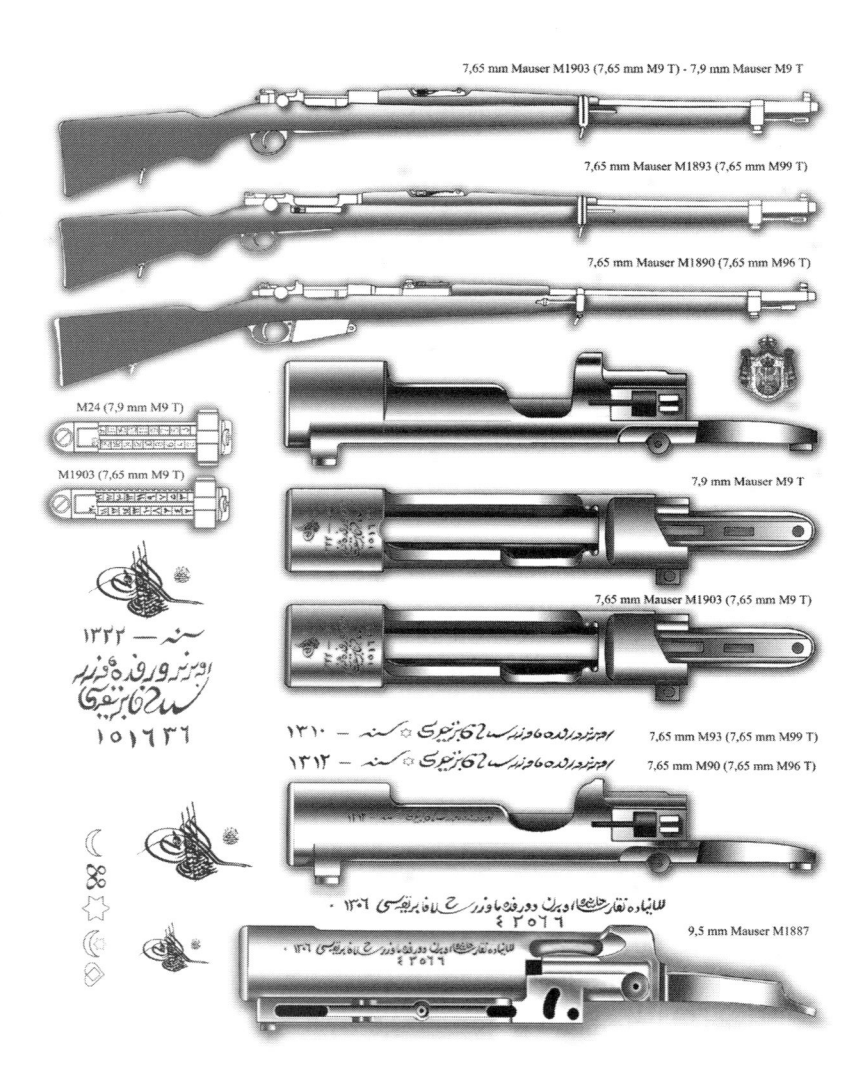

7,65 mm Mauser M1903 (7,65 mm M9 T) - 7,9 mm Mauser M9 T

7,65 mm Mauser M1893 (7,65 mm M99 T)

7,65 mm Mauser M1890 (7,65 mm M96 T)

M24 (7,9 mm M9 T)

M1903 (7,65 mm M9 T)

7,9 mm Mauser M9 T

7,65 mm Mauser M1903 (7,65 mm M9 T)

7,65 mm M93 (7,65 mm M99 T)

7,65 mm M90 (7,65 mm M96 T)

9,5 mm Mauser M1887

Figure 3-1. Top to bottom: Turkish Model 1903 converted to 7.92 x 57 mm calibre (M9T); Turkish Model 1893 7.65 x 53 mm (7.65 mm M99T) and Turkish Model 1890 7.65 x 53 mm (7.65 mm M96T), both models captured in the 1912 Balkan War. Also shown, the receiver of a converted M9T 7.92 x 57 mm; the receiver of an original Model 1903 7.65 x 57 mm and last, the receiver of an original Turkish Model 1887 9.5 x 60R mm rifle.

Serbian and Yugoslav Mausers

European countries as a sporting cartridge. The bullet size in the military cartridge is actually 7.92 mm (0.312 inch) but sporting cartridge bullets may range from 7.90 mm (0.311 inch) to 7.92 mm (0.312 inch).

The Turkish Model 1890 differed from the Belgian, and other Model 1889 rifles in that it had a short upper handguard that replaced the sheet steel tube surrounding the barrel. The barrel had a stepped contour with matching clearance cuts in the stock instead of a tapered barrel to eliminate bedding problems caused by the expansion of a hot barrel. The rear sight was mounted on a ring that encircled the base. A one-piece sear was employed in the trigger mechanism.

TURKISH MODEL 1893 MAUSER RIFLE

By the time Peter Paul Mauser made the last delivery of the Model 1890 7.65 x 53 mm rifles and carbines to Istanbul in 1893, he had developed and commercially marketed a new type of box magazine. First used on the Spanish Model 1893, the new magazine was enclosed entirely within the stock. It was loaded with cartridges held in a clip, a strip of metal folded to capture the rims of the cartridges, refer to Figure 3-1.

The shooter inserted the end of the clip in a slot in the rear receiver ring and pushed the cartridges down into the magazine where they stacked themselves in a zigzag arrangement. In addition to being quicker to reload, the cartridges were contained in a magazine at the rifle's center point and so the balance of the rifle did not change as the magazine emptied, as it did in rifles with under-the-barrel, tubular magazines.

The Porte immediately ordered 154,000 Model 1893 Mauser rifles also chambered for the 7.62 x 53 mm cartridge. The Turkish model was fitted with a magazine cutoff, the only Mauser (except for experimental designs) to be so equipped. The magazine cutoff made it possible to load the rifle one cartridge at a time while retaining a full magazine in reserve.

TURKISH MODEL 1903 7.65 X 53 MM MAUSER RIFLE

In 1898, Mauser completed his series of bolt-action, repeating arms with his most famous design, the Model 1898. The receiver was made larger to allow space for the new bolt that now had three locking lugs at the front. An improved safety was also incorporated, refer to Figure 3-1.

Pleased with its previous Mausers, Turkey signed a fourth contract for 200,000 rifles of the Model 1898 pattern, but chambered for the 7.65 x 53 mm cartridge instead of the German 7.92 x 57 mm cartridge. Six

Serbian and Yugoslav Mausers

years later, Turkey ordered another 7,617 Model 1903 rifles for its gendarmerie in Macedonia.

MARKINGS

Turkish Mausers were stamped with the Mauser company name in Turkish, the Sultan's name, "Abduelhamid II" (reigned 1876-1909), and the year of manufacture according to the Islamic calendar rather than the model year. The "Model 1887" was thus marked with the year "1306" for 1890, the "Model 1890" carried the dates "1309" or "1310" for 1893 and 1894, the "Model 1893" carried the date "1312" for 1896, and the "Model 1903" was dated "1322," "1323," "1324," and "1325" for 1906, 1907, 1908, and 1909, respectively. At the time of its decision to adopt a uniform calibre for its weapons, the Serbian Ministry of the Army decided that the captured Turkish weapons would be converted to 7.92 x 57 mm calibre but would not be converted into carbines. The Turkish Model 96Ts, Model 99Ts, and Model 9Ts were converted at the Military Technical Institute using new barrels and rear sights. The upper band and the original Turkish markings on the rifles were left intact, see Figure 3-2. The original Turkish bayonets were retained and reissued with the rifles.

Figure 3-2. Receiver markings, top: Converted Turkish Model 1903 7.92 x 57 mm Mauser dated in Arabic "1322" (1906) and serial numbered 160,147; Turkish Model 1903 7.65 x 53 mm converted to the Serbian M9T, showing in Arabic, the year "1322" (1906) and the serial number 105,636.

TURKISH ARMS AND SERBIA

Serbia became involved with the Turkish arms during the First Balkan War (1912), when the Serbian army seized a large number of Turkish rifles. These were immediately reissued to the Serbian infantry. A team of specialists from the Military Technical Institute at Kragujevac inspected

Serbian and Yugoslav Mausers

these rifles, and on 20 June 1914, they reported that ". . . the Turkish repeating rifle with straight line detachable box magazine (Model 1890) is valued at 55 dinars ($30.25), the Turkish repeater with cut-off magazine (the Model 1893) is valued at 65 dinars ($35.75), and the Turkish (Model 1903) repeating rifle (resembling the Serbian Model 1910) is [also] worth 65 dinars ($35.75)." The experts also valued the "old Turkish ten shot

Figure 3-3. Kragujevac armorers inspect Turkish Model 1893 7.65 x 53 mm Mauser rifles captured in the 1912 Balkan War.

rifle with tubular magazine in calibre of 9.5 mm," the Model 1887, as worth just 15 dinars ($8.25) each, see Figure 3-3. Most eventually became part of the Serbian Army's war reserve and were remarked as follows:

> The Model 1890 was marked "Model 96T" (T for Turkey)
> Model 1893 was marked "M 99T"
> Model 1903 was marked "Model 9T"

This marking system is believed to have originated from the Arabic language signatures on the Turkish rifle that listed model dates in accordance with the Islamic calendar.

CHAPTER 4
THE KINGDOM OF THE SERBS, CROATS, AND SLOVENES: 1918-1929
THE KINGDOM OF YUGOSLAVIA: 1929-1941

THE MODEL 1924 MAUSER RIFLE

The Kingdom of the Serbs, Croats, and Slovenes (SHS), the predecessor of Yugoslavia, was a new state established in 1918 in the wake of World War I. Its army inherited twenty-six models of rifles in six different calibres. An inventory dated 11 February 1919 indicated that the Army had 72,183 French Mle. 07/15 Lebel 8 mm rifles; 552 Mle. 1890 Lebel 8 mm carbines and Mle. 1892 8 mm musketoons, and about 10,000 M1899, M1899/07, M1910, and M1880/07 Serbian Mauser rifles in 7 x 57 mm calibre (repaired in France in 1916).

When the French occupation forces evacuated Hungary in September 1919, they turned over more French rifles to the SHS Army including more than 14,000 Mle.1886/93, Mle. 07/15, and Mle.16 Lebel 8 mm rifles and 5,800 Mle.1890, Mle.07/15 and Mle.16 Lebel 8 mm carbines originally from the "Armeé de Hongrie."

After the Versailles Peace resolutions were starting in July 1920, the Army of the Kingdom of Serbs, Croats, and Slovenes then received 500 railroad cars of armament and war equipment from Austria. From the mid- to late 1920s, the SHS received an estimated total of 1,500 railroad cars of rifles collected from the battlefields and from the civilian population amounting to nearly 75,000 more rifles of Austrian, Italian, Russian, and German origin. This brought in about 150,000 more Austrian 8 mm Mannlicher Model 88/90, Model 90, and Model 95 rifles, and 27,000 Austrian 8 mm Mannlicher Model 90 and Model 95 carbines, 11,000 Italian 6.5 mm Mannlicher-Carcano M1891 rifles, 34,000 Russian 7.62 mm Mosin-Nagant M1891 rifles, and 1,600 German 7.92 mm Mauser M1898 rifles.

The SHS Army hoped to acquire new rifles from the new Republic of Czechoslovakia (RCS). But a faction in the Army that had developed close ties to the French military insisted on acquiring French small arms, even though they had been shown to be inferior to German arms. A bitter debate ensued in the SHS parliament that was ultimately resolved in favor of the Czech Model 98/22 in 7.92 mm calibre. The decision was confirmed on 22 March 1923 but not implemented.

Serbian and Yugoslav Mausers

In the meantime, the English firm of Birmingham Small Arms (BSA) offered the SHS 500,000 surplus Lee-Enfield SMLE No. 1 Mk. III and No. 1 Mk. III* rifles. Ultimately, the offer faltered over financing.

The Belgian Connection
In 1925, with the problem still unresolved, the SHS government agreed to make funding available to purchase new small arms but insisted on the Mauser Mle. 1924 rifle as manufactured by Fabrique Nationale in Herstal-lèz-Liège, Belgium, see Figure 4-1.

A total of 100,000 Mauser Mle. 1924 rifles, 110,000,000 rounds of 7.92 x 57 mm ammunition plus machinery and other equipment to manufacture the rifles in the SHS were finally agreed upon. Equipment installation and training took place from 1925 to 1927 as the complete FN production line was duplicated at Kragujevac. Prototype testing was completed in December 1927 and the first production rifles rolled off the production line, see Figure 4-2.

Full production began on 28 October 1928. The factory's 1928 production barely exceeded thirty rifles a day, but eventually reached 200 rifles and 200,000 cartridges per day. Delays were due primarily to acquiring the specified steels, all of which had to be imported—a problem that plagued many gun manufacturers then, as well as today.

Figure 4-1. Rifle production at the Fabrique Nationale d'Armes de Guerre, Herstal-lèz-Liège, in 1912.

Serbian and Yugoslav Mausers

By the start of the Second World War, Yugoslavia's Mauser production had provided the Army approximately one million Model 24 7.92 x 57 mm rifles, see Figure 4-3. This total did not include the Turkish, Serbian, German, and Mexican pattern Mausers in Yugoslav inventory, as well as the Austrian Mannlichers that had also been altered to 7.92 x 57 mm calibre. Suffice it to say, Kragujevac had mastered the production of the Model 24 7.92 x 57 mm rifle and carbine.

Figure 4-2. V Department, VTZ, at Kragujevac in 1928.

Model 1924 Rifle and Carbine Details

Both rifles and carbines were manufactured and the differences are in the details, see Figure 4-4. The carbine featured in the two versions: **Type 1**, with a turned-down bolt handle for cavalry in contrast to the rifle's straight bolt handle, and **Type 2** with a straight bolt handle, for infantry and technical troops, bicyclist's platoon, etc., see Figures 4-5 and 4-6. The carbine's lower band had two sling swivels, one placed beneath and the other on the left side. The two lower sling swivels were located on the bottom of the buttstock, one forward of the butt plate, and the other on the left side, just behind of the stock wrist. The twin sling

Figure 4-3. Serbian army private armed with Model 1924 7.92 x 57 mm infantry rifle. He is wearing the Model 1895 ammunition pouches.

Serbian and Yugoslav Mausers

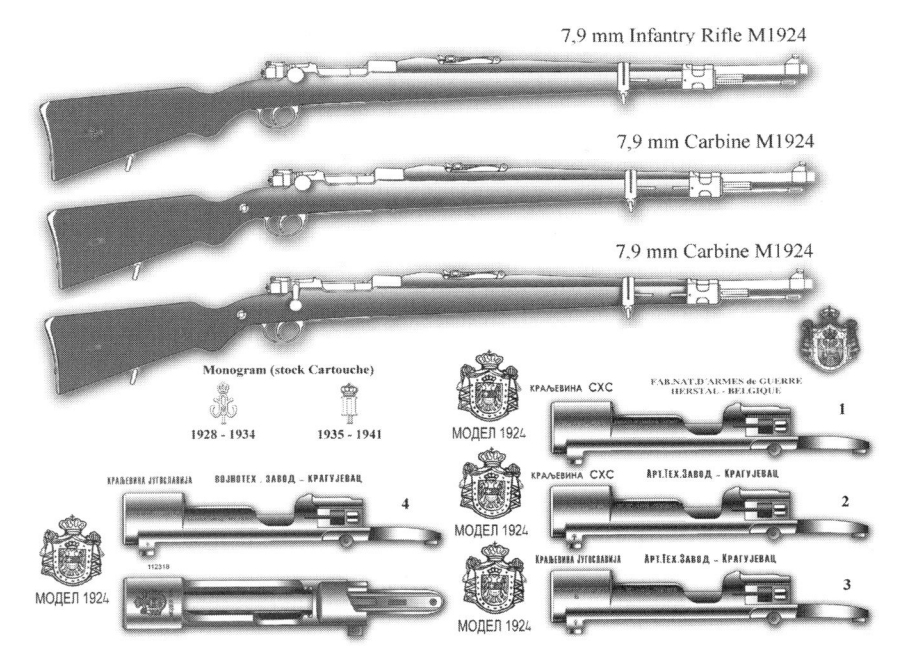

Figure 4-4. Top to bottom: Model M1924 7.92 x 57 mm Mauser infantry rifle; the Model 1924 7.92 x 57 mm Type II carbine; the Model 1924 7.92 x 57 mm Type I carbine; (1) Belgian-made receiver, 1926; (2) ATZ, 1928-1929; (3) ATZ 1929-1931; (4) VTZ, 1931-1941.

swivels enabled mounted troops to carry the carbine over the shoulder with the butt secured to the waist belt by a special attaching strap when mounted, or slung over the shoulder when on foot, refer to Figure 4-5 for details of the markings and sling swivel mountings.

Dimensions

The Mauser Model 1924 was an "intermediate-ring" model, see Figure 4-7. It was basically a Model 1898 Mauser rifle as developed for the German armed forces, and which had proven superior to most Allied infantry rifles during World War I with the exception of the American Model 1903 Springfield (a Mauser 98 derivative) and the British/American Pattern 1914/Model 1917 Enfield. The Model 1924 employed the same five-round magazine with a removable floor plate contained with the stock, the same bolt (see Figure 4-8), safety and trigger mechanism (see Figure 4-9) as the Model 1898 7.92 x 57 mm Mauser.

Serbian and Yugoslav Mausers

Figure 4-5. Details of the Model 1924 carbine. Clockwise from top left: carbine sling swivel on buttstock; sling attaching point on forward barrel band; flattened, knurled bolt handle; receiver markings; shape of carbine bolt handle.

There are three basic Mauser rifle actions: "large-ring," "small-ring," and "intermediate-ring." "Ring" refers to the receiver ring.

The Model 1898's actions are typically called the **large-ring** models and the pre-Model 1898, the **small-ring** models (the Models of 1893, 1894, 1895 and 1896). The **large-ring** Model 1898s could handle far more pressure than the small-ring models. Large-ring Mausers also cock on opening the bolt while the small-ring Mausers cock on closing. The Model 1898 with its large 35.8 mm (1.41-inch) diameter ring also had better gas-venting ability if a cartridge case ruptured.

The term "large-ring" also refers to a standard length. The distance between floor plate screw holes, center to center, in the large-ring Mauser is 199 mm (7.835 inches) and the overall length of the action is 222.25 mm (8.75 inches). The large-ring class typically includes the Mauser Model 1898 in all its configurations: G98, K98, K98k, etc. It also in-

Figure 4-6. Two types of bolt handles were used on the M1924 Cavalry Carbine. Above: Type I turned-down bolt handle. Below: Type II straight bolt handle.

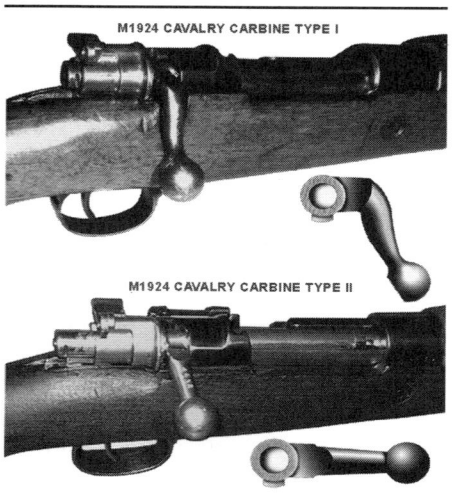

Serbian and Yugoslav Mausers

Figure 4-7. A Model 1924 Infantry rifle and Model 1924CK carbine manufactured through 1941; details shown clockwise from top, left: left side of receiver with manufacturer's marking; receiver ring showing national crest and model markings in Serbian; (infantry rifle and carbine) serial number; receiver ring shows the national crest and Model 1924CK; front barrel band and bayonet mount detail.

cludes the Czech vz.24, vz.98/22, Mexican Model 1912, Argentine Model 1909, and so on.

The **small-ring** Mauser action is 217.4 mm (8.56 inches) in length. The distance between floor plate screw holes is 193.5 mm (7.62 inches) center to center and the receiver ring diameter is 33 mm (1.299 inches). The small-ring Mausers include the Serbian Model 1899, Model 1899/07 and the Swedish Models of 1894 and Model 1896, among others.

The **intermediate-ring** models share many of the same measurements with the small-ring models except that the front ring is 35.8 mm

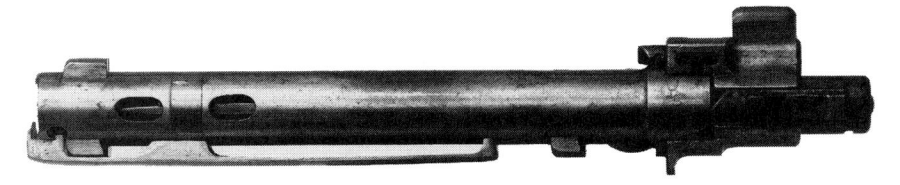

Figure 4-8. Model 1924 bolt assembly

Serbian and Yugoslav Mausers

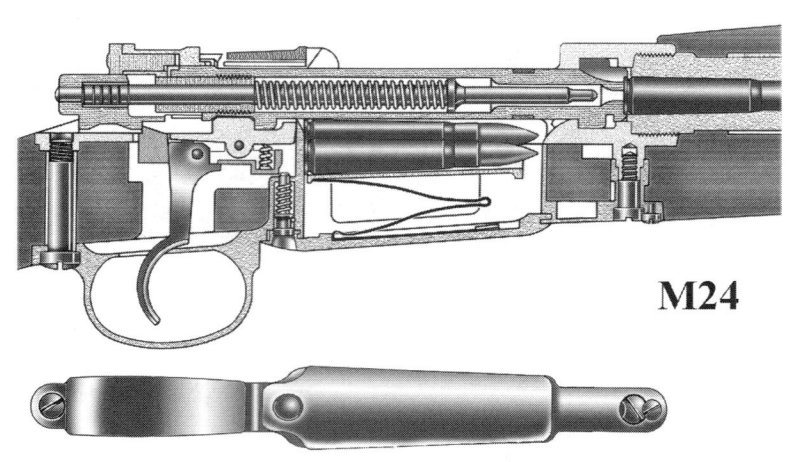

Figure 4-9. Cross section of the Model 1924 rifle receiver.

Figure 4-10. Receiver markings on the Model 1924, Model 1924CK and Model 1924B rifles: 1-3, National crest and model designation; 4) the factory and country name (Kraljevina SHS or Kraljevina Jugoslavija = Kingdom of Serbs, Croats and Slovenes or Kingdom of Yugoslavia); factory markings (5) FN, (6) ATZ, (7) VTZ.

Serbian and Yugoslav Mausers

(1.41 inches), the same size as the Model 1898 large receiver ring. But the intermediate-ring Mauser is 6.35 mm (0.25 inch) shorter than large-ring actions at 215.9 mm (8.5 inches) in length. Floor plate screw holes are 193.5 mm (7.62 inches) center to center as in the small-ring actions. The intermediate-ring models include all actions designated as the Model 1924 such as the FN Model 1924 and Yugoslavian Model 1924. The single exception is the Czech vz.24.

The overall length of both carbines and infantry rifles was the same, 1,094 mm (43.1 inches) with barrels 504 mm (19.8 inches) long. Both the carbine and rifle had rear sights adjustable in elevation from 200 to 2,000 meters (218.7 to 2,187.2 yards). Both the rifle and carbine barrels had four grooves with a right twist of one turn in 240 mm (1:9.4 inches). Muzzle velocity was 730 meters (2,395 feet) per second. See Table 4-1.

Table 4-1 Technical Data, Model 1924 Mauser Rifle	
Calibre	7.92 x 57 mm
Rifling	concentric, 4 grooves, 0.13-0.18 mm deep and 4.2-4.4 mm wide; 1 turn in 240 mm (9.4 inches), right hand (pitch 5° 54')
Bolt	3 locking lugs
Magazine	internal staggered-column box, capacity 5 rounds
Loading system	charger, or single rounds
Length overall	1,094 mm (+2, -3) 43.1 inches
Barrel length	504 mm 19.8 inches
Weight	3,850 gm (+200, -170) 8.55 lbs
Sights	(front) unprotected barleycorn; (rear) a tangent leaf graduated from 200 to 2000 m (218 to 2,187 yds)
Action	Intermediate-ring Mauser 35.8 mm (1.41 inches) in diameter and 215.9 mm (8.5 inches) long. Screw spacing 193.5 mm (7.62 inches). Bolt length 155.3 mm (6.115 inches). Magazine length 820.9 mm (3.232 inches).

Serbian and Yugoslav Mausers

Markings and Finishes

Markings and finishes for the Model 1924 are listed in Table 4-2. Both the rifle and carbine were stamped "Model 1924" on the receiver ring below the national coat of arms. Stamped on the left side of the receiver in Cyrillic was the name of the manufacturer, "*Kraljevina SHS*" (for

Table 4-2
Serial Number Ranges and Markings
Model 1924 Rifles and Carbines

Year	Total Number Per Year		Serial # (Start/End)	Markings on Receiver
	M1924	M1924CK		
1926-28	100,000		А1-А10,000; Б10,001-Б20,000; В20,001-В30,000; Г30,001-Г40,000; Л40,001-Л50,000; Ђ50,001-Ђ60,001; Е60,001-Е70,000; Ж70,001-80,000; 380,001-390,000; И90,001-И100,000	Fab.Nat. D'Armes de Guerre/ Herstal-Belgique **КРАЉЕИНА СХС**
25 Oct. 1928-31 Dec. 1929	1,860		(A)1-(A)1,860	АРТ . ТЕХ . ЗАВОД КРАГУЈЕАЦ **КРАЉЕВИНА СХС**
1929	56,800		(A)1,861-(A)58,660	АРТ . ТЕХ . ЗАВОД КРАГУЈЕАЦ **КРАЉЕВИНА СХС**
1930	56,000		(A)58,661-(A)99,000 100,000-115,460	АРТ . ТЕХ . ЗАВОД КРАГУЈЕАЦ **КРАЉЕВИНА ЈУГОСЛАВИЈА**
1931	56,800		115,461-172,260	АРТ . ТЕХ . ЗАВОД КРАГУЈЕАЦ **КРАЉЕВИНА ЈУГОСЛАВИЈА**
1932	56,900		172,261-229,060	ВОЈНТЕХ . ЗАВОД - КРАГУЈЕВАЦ **КРАЉЕВИНА ЈУГОСЛАВИЈА**
1933	56,800		229,061-285,860	ВОЈНТЕХ . ЗАВОД - КРАГУЈЕВАЦ **КРАЉЕВИНА ЈУГОСЛАВИЈА**
1934	56,800		285,861-342,660	ВОЈНТЕХ . ЗАВОД - КРАГУЈЕВАЦ **КРАЉЕВИНА ЈУГОСЛАВИЈА**
1935	56,800	6,000 (?); known from #223 to 1,822	01-56,800	ВОЈНТЕХ . ЗАВОД - КРАГУЈЕВАЦ **КРАЉЕВИНА ЈУГОСЛАВИЈА**
1936	56,800		56,801-113,600	ВОЈНТЕХ . ЗАВОД - КРАГУЈЕВАЦ **КРАЉЕВИНА ЈУГОСЛАВИЈА**
1937	56,800		113,601-170,400	ВОЈНТЕХ . ЗАВОД - КРАГУЈЕВАЦ **КРАЉЕВИНА ЈУГОСЛАВИЈА**
1938	56,900		170,401-227,200	ВОЈНТЕХ . ЗАВОД - КРАГУЈЕВАЦ **КРАЉЕВИНА ЈУГОСЛАВИЈА**
1939	56,800		227,201-284,000	ВОЈНТЕХ . ЗАВОД - КРАГУЈЕВАЦ **КРАЉЕВИНА ЈУГОСЛАВИЈА**
1940	56,800		284,001-340,000	ВОЈНТЕХ . ЗАВОД - КРАГУЈЕВАЦ **КРАЉЕВИНА ЈУГОСЛАВИЈА**

Serbian and Yugoslav Mausers

Year	Total Number Per Year		Serial # (Start/End)	Markings on Receiver
	M1924	M1924CK		
1941	?	6,000 (?); known from #223 to 1,822	340,001	ВОЈНТЕХ . ЗАВОД - КРАГУЈЕВАЦ КРАЉЕВИНА ЈУГОСЛАВИЈА

Table 4-2, cont.
Serial Number Ranges and Markings
Model 1924 Rifles and Carbines

M24 Infantry rifle: 2 sling swivels on buttstock, narrow lower band, straight bolt.
M24 Carbine: 4 sling swivels (2 on barrel band, 2 on buttstock), straight or curved bolt.
M24CK *Chetnik's* rifle: 2 sling swivels on buttstock, narrow lower band behind "H" band, curved bolt.

the Kingdom of Serbs, Croats, and Slovenes) from 1924 to 1929, and "*Kraljevina Jugoslavija*" (for the Kingdom of Yugoslavia) after 1929.

The ruler's cypher from 1926 to 1934 was a "crown" over a stylized Cyrillic letter "A" over the Roman numeral I for King Alexander I. From 1935 to 1941, the ruler's cypher was a "crown" over the Cyrillic letter "П" over the Roman numeral II for King Peter II.

The manufacturer's name was stamped on the left side of the receiver. The first 100,000 Model 1924 rifles were manufactured in Belgium and were marked in French:

FAB. NAT.D'ARMES de GUERRE/HERSTAL – BELGIQUE

See Figure 4-10 for the Cyrillic markings stamped on the receiver.
Rifles and carbines manufactured in the SHS and later Yugoslavia between 1928 and 1931 bore the name of the Kragujevac plant in Cyrillic:

ART. TEH. ZAVOD KRAGUJEVAC
(Artillery Technical Institute, Kragujevac)

After 1931, the marking was changed to

VOJ. TEH. ZAVOD KRAGUJEVAC
(Military Technical Institute, Kragujevac)

again in Cyrillic, refer to Figure 4-10.

Serbian and Yugoslav Mausers

Every weapon was proofed and stamped with the smokeless powder proof mark, the crowned "T," along with the inspector's mark and serial number. Serial numbers on the 100,000 rifles made in Belgium between 1928 and 1930 were in Arabic numerals but preceded by the Cyrillic letter "A" to denote the first batch of 99,000 rifles. Succeeding Kragujevac-made weapons had serial numbers in Arabic numerals but without a preceding letter.

It bears repeating that all rifles and carbines manufactured during the period 1926 to 1934 were marked on the buttstock with the royal cipher of King Alexander I Karageorgevich (1888-1934). For weapons made between 1935 and 1941, this mark was replaced by the cipher, or crest, of King Peter II Karageorgevich as mentioned above. Also, refer to Figure 4-7. See Table 4-3 for a complete list of markings and Table 4-4 for the type of finishes applied to the major parts of all Model 1924 rifles.

The Model 1924 7.92 x 57 mm Mauser rifle and carbine became the backbone of the pre-World War II Yugoslav Army. They were issued to all military and paramilitary troops, see Figures 4-11 and 4-12.

Table 4-3
Markings, Model 1924, Model 1924CK and Carbine Model 1924, 7.92 mm Rifles

	See Also Table C-1, C-2, C-3, C-4, C-5, C-6, C-7, C-8	Coat of Arms and Pattern Mark	Manufacturer's Markings	Country's Name	Year of Production	Ruler's Monogram	Full Serial Number	Last Three or Four Digits of Serial Number	Smokeless Powder Proof Mark	Inspection Mark	Acceptance Mark
Barrel:							x		x	x	x
Receiver Ring:											
Top	x										
Left			x						x	x	
Right						x					
Receiver											
Left			x								

Serbian and Yugoslav Mausers

Table 4-3, cont.
Markings,
Model 1924, Model 1924CK and Carbine Model 1924, 7.92 mm Rifles

See Also Table C-1, C-2, C-3, C-4, C-5, C-6, C-7, C-8	Coat of Arms and Pattern Mark	Manufacturer's Markings	Country's Name	Year of Production	Ruler's Monogram	Full Serial Number	Last Three or Four Digits of Serial Number	Smokeless Powder Proof Mark	Inspection Mark	Acceptance Mark
Right										
Recoil Lug									x	
Bolt Assembly										
Bolt Body										
Bolt Handle Base									x	
Bolt Handle Neck						x				
Bolt Handle Ball								x		
Firing Pin										
Cocking Piece									x	
Bolt Sleeve with Gas Shield							x		x	
Bolt Sleeve Stop									x	
Safety Lever							x		x	

Serbian and Yugoslav Mausers

Table 4-3, cont.
Markings,
Model 1924, Model 1924CK and Carbine Model 1924, 7.92 mm Rifles

See Also Table C-1, C-2, C-3, C-4, C-5, C-6, C-7, C-8	Coat of Arms and Pattern Mark	Manufacturer's Markings	Country's Name	Year of Production	Ruler's Monogram	Full Serial Number	Last Three or Four Digits of Serial Number	Smokeless Powder Proof Mark	Inspection Mark	Acceptance Mark
Extractor									x	
Extractor Collar									x	
Ejector									x	
Bolt Stop									x	
Trigger Assembly										
Trigger									x	
Sear-fork									x	
Trigger Guard Plate						x				
Magazine Floor Plate						x			x	
Stock				x					x	x
Butt Plate					x					
Front Band, Lower Band									x	

Serbian and Yugoslav Mausers

Figure 4-11. Serbian army infantry platoon equipped with the 7.92 x 57 mm Model 1924 rifles and converted Light Machine Gun 7.92 x 57 mm CSRG M15/26. The enlisted men are wearing the Yugoslavian M1930 ammunition pouches.

Figure 4-12. Serbian army bicycle platoon equipped with the Model 1924 Cavalry carbine.

Serbian and Yugoslav Mausers

Table 4-4
Finishes,
M1924, M1924CK, M1924B Rifle and Carbine

Part	Finish
All wood parts	Sanded smooth and oiled
Barrel	Blued, used a hot salt bath method after polishing and degreasing
Receiver	Blued, used a hot salt bath method after polishing and degreasing
Front band, Bayonet Mount, Lower Band	Blued, used a hot salt bath method after polishing and degreasing
Metal plate on the underside of the butt	Blued, used a hot salt bath method after polishing and degreasing
Swivels	Blued
Butt Plate	Finished in-the-white
Recoil Lug	Blued by heating
Front Sight Blade	Blued by heating
Rear Sight Base	Blued, used a hot salt bath method after polishing and degreasing
Sight Leaf	Blued
Slide	Blued
Slide Catch	Blued
Screws	Finished with a black coating
Bolt Parts	Finished in-the-white
Bolt Stop	Blued by heating
Bolt Sleeve Lock Plunger	Blued by heating
Trigger Guard and Floor Plate	Blued, used a hot salt bath method after polishing and degreasing
Trigger	Blued
Sear	Finished in-the-white
Follower	Blued

Serbian and Yugoslav Mausers

Table 4-4, cont. Finishes, M1924, M1924CK, M1924B Rifle and Carbine	
Part	Finish
Front and Rear Trigger Guard Screws, Front Lock Screw	Blued
Trigger Guard Screw Tube	Finished in-the-white
Magazine Spring	Blued spring steel
Floor Plate Catch	Blued by heating

THE *CHETNIK* MODEL 1924 ASSAULT RIFLE
At the beginning of the Second World War, the Military Technical Institute's Kragujevac Factory designed and perfected what became known as the "*Chetnik*" Model 1924 assault rifle. It was intended for the special assault troops of the Yugoslavian Army, see Figure 4-13.

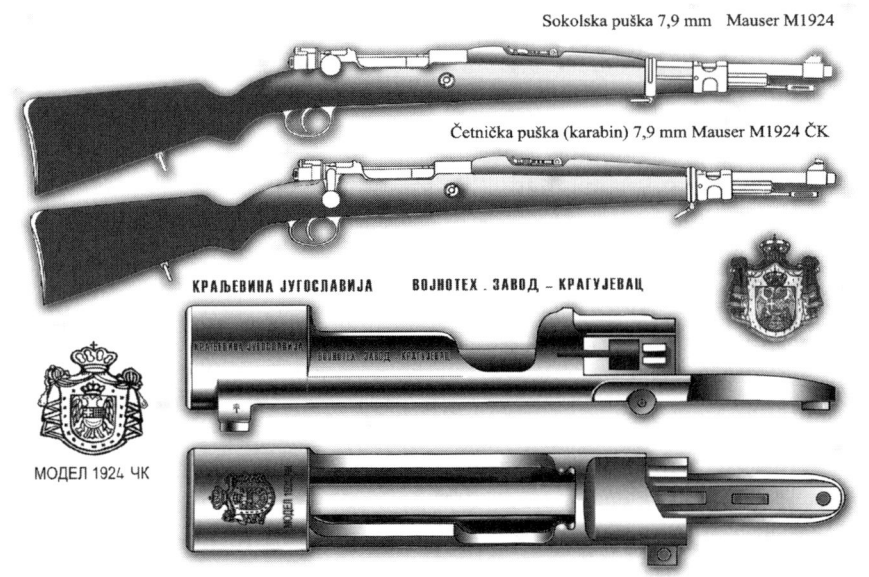

Figure 4-13. (Above) Serbian Sokol rifle Model 1924 7.92 x 57 mm; (below) Model 1924CK *Chetnik* (assault) rifle.

Serbian and Yugoslav Mausers

NOTE: The term *Chetnik* is used as a synonym for Yugoslav special operations troops; it derives from the Turkish term "çete" denoting guerrilla units.

The *Chetnik* rifle evolved from a combination of the original Fabrique Nationale Model 1924 carbine, which had an overall length of 940 mm (37 inches) with a barrel length of 405 mm (15.94 inches), the Czechoslovakian short Gendarme rifle Model 1916/33 carbine, which was 995 mm (39.2 inches) long, and the Iranian musketoon Model 1898/29, which was 967 mm (38.1 inches) long and with a barrel length of 455 mm (17.92 inches).

The "*Chetnik*" or "assault rifle" resulted from the formation of a new type of infantry which was established within the Yugoslavian Army on 26 April 1940 by the Minister of the Army and Navy. The *Chetnik* command was approved on 10 July 1940 by Milan Nedich, the Minister of War who renamed the *Chetnik* units as Assault units. Consequently, the newly formed *Chetnik* Command and its battalions changed their designations to Assault Command and Assault Battalions, respectively. The "Assault" designation failed to catch on and on 1 April 1941, five days before the war with Germany began, the Ministry of the Army and Navy renamed the Assault Command Headquarters as *Chetnik* Command Headquarters.

The *Chetnik* establishment consisted of a headquarters element and six battalions totaling fourteen companies and six communication platoons. Each battalion in turn consisted of a headquarters element and from two to three companies made up of three platoons, each with 70 privates. Battalion headquarters also had a communication platoon of 30 privates. *Chetnik* battalions had a total of from 450 to 660 combatants.

Production of assault rifles at the Military Technical Institute at Kragujevac did not start until May 1932. Based on the number of men authorized for the six battalions and headquarters elements, the number of rifles manufactured did not exceed 6,000 rifles. Unfortunately, surviving assault rifles found to date are only known in a serial number range of from 223 to 1,822.

NOTE: It is thought by most authorities that the Wehrmacht seized the majority of the Model 1924CK rifles and transported them to Germany. The serial numbers cited belong to the weapons preserved in Yugoslavia and thus the discrepancy in serial numbers.

Serbian and Yugoslav Mausers

Dimensions
The *Chetnik*, or assault, rifle's overall length measured 955 mm (37.6 inches) with a barrel length of 455 mm (17.92 inches). It featured the 2 sling swivels under the stock, a narrow lower band installed directly against the "H-shaped" upper band and a fixed rear sight that was adjustable in elevation from 200 to 1,000 meters (218 to 1,094 yards).

Markings and Finishes
The *Chetnik*, or assault, rifle's receiver ring was marked with the Yugoslavian national coat of arms followed by the model designation in Cyrillic, Model 1924CK. The designation "C" stood for *Chetnik* and the "K" for carbine. Refer to Table 4-3 and Figures 4-7 and 4-13 for a comparison of the Model 1924 infantry, carbine and *Chetnik* rifle receiver markings. On the left side of the receiver was stamped in Cyrillic:

> KRALJEVINA JUGOSLAVIJA
> (Kingdom of Yugoslavia)

and the manufacturer's name:

> VOJ. TEH. ZAVOD, KRAGUJEVAC
> (Military Technical Institute, Kragujevac)

By a decree dated 24 April 1941, soldiers assigned to *Chetnik* units received a distinctive uniform modeled after that of the mountain troops. The decree also described the *Chetnik* or assault knife and its black parade knot with the skull and crossbones (12 x 15 mm/0.472 x 0.590 inch) in size.

AMMUNITION AND AMMUNITION CARRIERS FOR THE MODEL 1924 MAUSER
The Model 1924 Mauser rifle in SHS/Yugoslavian service was chambered for the 7.92 x 57 mm Mauser cartridge. Initial supplies were obtained from Belgium but facilities were developed at Kragujevac to manufacture cartridges domestically, see Figure 4-14. This was the same cartridge developed for, and used by the German military in both World Wars.

Ammunition was carried in stripper clips holding five cartridges, see Figure 4-15. To load the rifle, the bolt was drawn back, the end of the stripper clip inserted into the cartridge guide in the receiver and the cartridges. The cartridges were then pushed down into the magazine with the thumb.

Serbian and Yugoslav Mausers

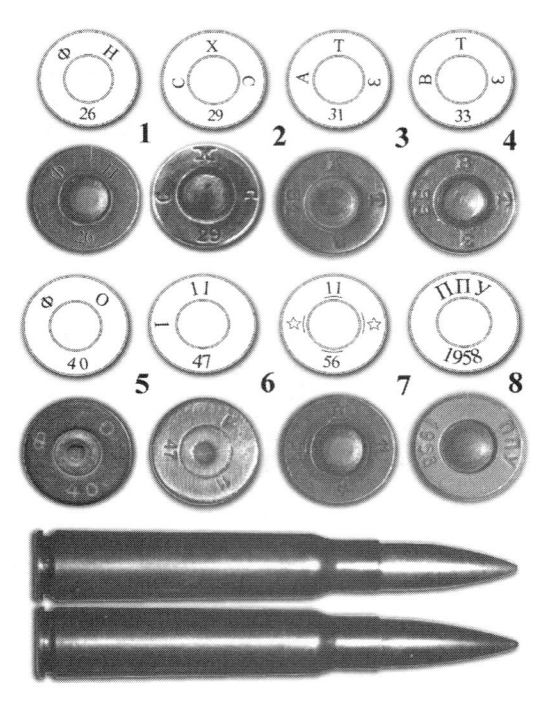

Figure 4-14. Yugoslav 7.92 x 57 mm ammunition. 1) Fabrique Nationale d'Armes de Guerre, Herstal-lèz-Liège (FN); 2) Kragujevac (SHS) 1928-29; 3) Kragujevac (ATZ) manufactured between 1930 and 1931; 4) Kragujevac (VTZ) manufactured between 1931-1941; 5) FOMU, Uzice (FO) manufactured 1940–1941; 6, 7) Uzice (11) manufactured 1948-1956; 8) Uzice (PPU), manufactured after 1956.

Ammunition in clips was carried in a leather ammunition pouch worn on the soldier's belt, see Figure 4-16. Depending on his type of service, he would wear one or two ammo pouches. Because elements of the SHS/Yugoslavian military were also equipped with Austrian Mannlicher rifles that had been converted to fire 7.92 mm cartridges, the cartridge boxes were "raked," or tilted to the left. This made it possible for the cartridge boxes to also hold the slanted cartridge clips used by the Mannlicher rifles.

The Model 1930 ammo pouch held a total of 30 cartridges in 6 clips of either the

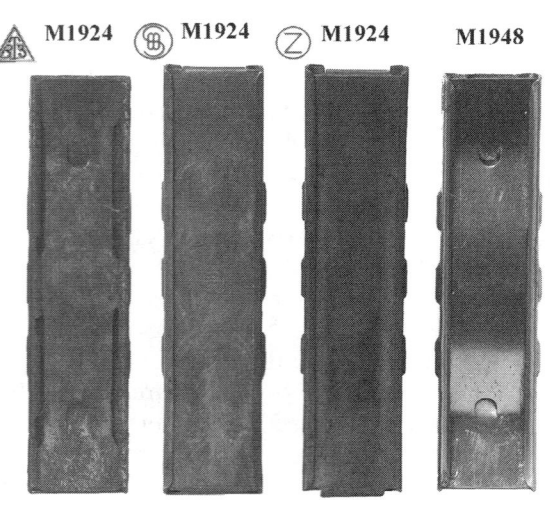

Figure 4-15. Ammunition clips for the Serbian Model 1924 ammunition.

102

Serbian and Yugoslav Mausers

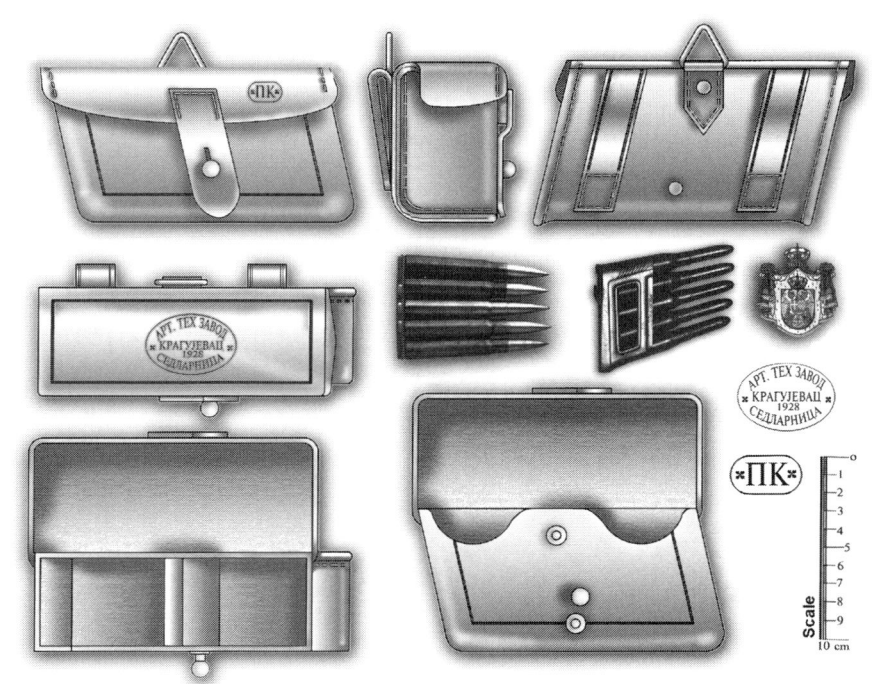

Figure 4-16. Yugoslav ammunition pouch, Model 1930.

Mauser or Mannlicher ammunition. The top of the pouch was marked in Cyrillic,

ART.TEH.ZAVOD KRAGUJEVAC/SEDLARNICA
(Artillery Technical Institute Kragujevac, Saddle Workshop)

followed by the year of production. The front of the pouch was stamped with an inspector`s mark in Cyrillic, "∗ПК∗" ("PK"=Received). The boxes were 50 mm (2 inches) wide by 140 mm (5.5 inches) long and 90 mm (3.5 inches) high.

Ammunition in clips was packed in the cardboard boxes, see Figure 4-17. A box held a total of 15 cartridges in 3 clips of 7.92 x 57 mm cartridges. The top of the box was marked in Cyrillic,

15. BOJ. METAKA za PUSKU MOD. 24. g. KALIBRA 7.9 MM
ZRNA DUPLO SPICASTA/UPRAVA ART. TEHN. ZAVODA

Serbian and Yugoslav Mausers

(15 cartridges for rifles Mod. 24 year, calibre 7.9 mm with a streamlined "boat-tail" pattern bullet/Artillery [or Military] Technical Institute)

The boxes were 46 mm (1.8 inches) wide by 65 mm (2.56 inches) long and 86 mm (3.4 inches) high.

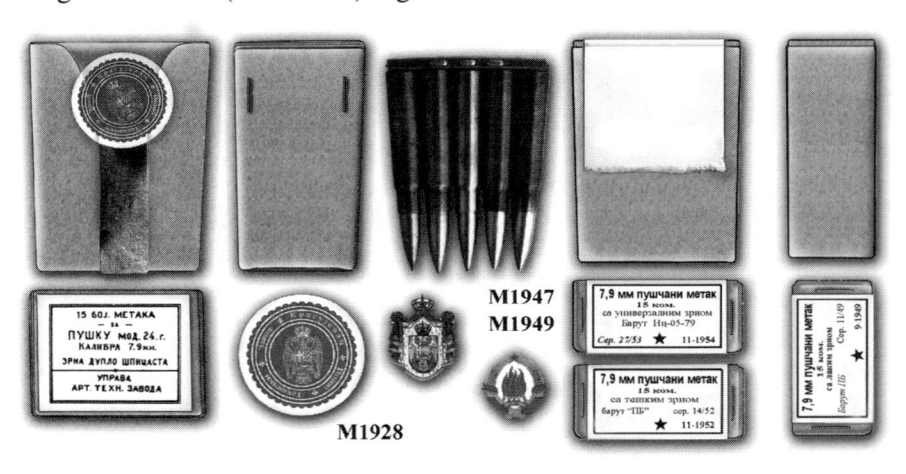

Figure 4-17. Yugoslav ammunition boxes, left to right: Model 1924; Model 1947 and Model 1949.

ACCESSORIES AND BAYONET

The Military Technical Institute at Kragujevac produced accessories and spare parts as well as the bayonets for the Model series rifles, see Figures 4-18 and 4-19.

The Model 1924 bayonets (see Figure 4-21) had a steel hilt fitted with two wooden grips, fixed by two screws set in washers. The right forward side of the cross-guard was marked either "ATZ" or "VTZ" (for "Artillery" or "Military Technical Institute, Kragujevac"). The single-edged blade was 380 mm (15 inches) long, with a fuller on each side. The scabbard was made of steel, see Figure 4-20. They were issued to the Army with Yugoslav Model 1924 leather belt frogs, see Figure 4-21.

The Model 1924CK bayonet featured what is known as a kindjal-shaped blade, a double-edged blade with the point at the center line and a fuller down the middle on both sides, refer to Figure 4-20 and see Figure 4-22 . The *Chetnik* unit's distinctive assault skull badge was attached to

Serbian and Yugoslav Mausers

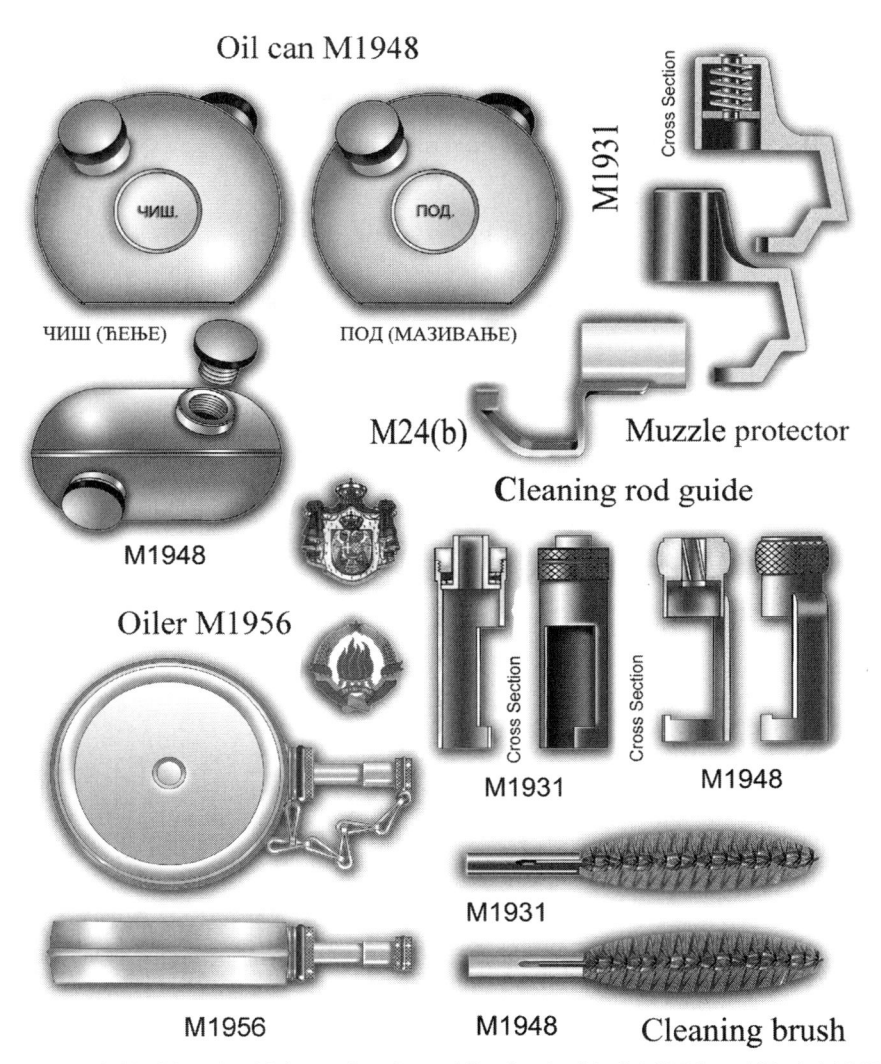

Oil can M1948

ЧИШ (ЋЕЊЕ)

ПОД (МАЗИВАЊЕ)

M1931

M24(b)

Muzzle protector

M1948

Cleaning rod guide

Oiler M1956

M1931

M1948

M1956

M1931

M1948

Cleaning brush

Figure 4-18. Cleaning kit issued to the soldier for the Model 1924 and Model 1948 rifles.

the metal scabbard and a sword knot with a 25 mm/0.9 inch long tuft was attached to the pommel. The *Chetniks* were also armed with brass-knuckles, and hand grenades.

Serbian and Yugoslav Mausers

Figure 4-19. Original package of spare parts; in this example, sears for the Model 1924 rifle.

The Sokol Rifle

Using the Model 1924CK assault rifle as its pattern, the Kragujevac factory also developed a special weapon intended for training men belonging to the Kingdom of Yugoslavia's "Sokol's Association."

NOTE: The "Sokol" (Falcon) was a large pre-World War II Serbian/Yugoslav youth organization that emphasized physical training and fitness and the development of outdoor survival skills.

The "Sokols" rifle differed from the Model 1924CK rifle in that it was 945 mm (37.2 inches) long compared to 955 mm (37.6 inches) with a straight bolt handle. The lower band was 76 mm (3 inches) behind the upper band which carried a single sling swivel on the bottom and no side-mounted sling swivel, refer to Figure 4-13. It could also fire a 5.4 mm (.22-calibre long rifle) rimfire cartridge for economy using a "Small Calibre Target Device."

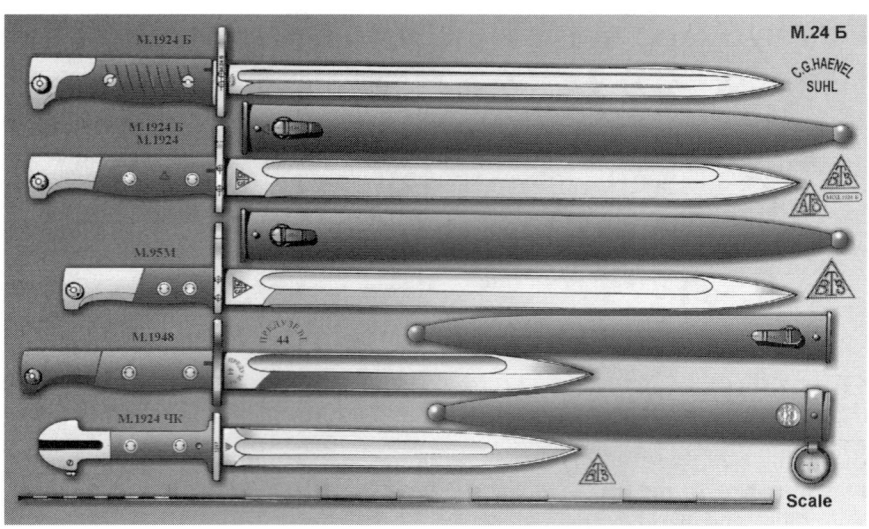

Figure 4-20. Bayonets, top to bottom: Model 1924B, Model 1924B and Model 1924, Model 1895M, Model 1948, and Model 1924CK.

Serbian and Yugoslav Mausers

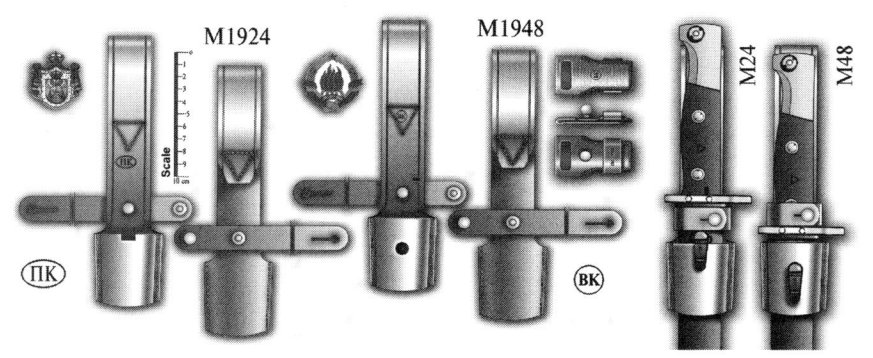

Figure 4-21. Bayonet frogs for the Model 1924 and Model 1948 rifles. Inset, buckles for the Model 1948 sling.

Lazar Jovanovich had submitted his design for the "Small Calibre Target Device" to the Yugoslav patent office, the Institute for Industrial Properties Protection, in July 1934, see Figures 4-23 and 4-24. His patent request shows that he had modified a standard barrel with a 5.4 mm (0.212-inch) barrel insert and a bolt adapter, see Figures 4-25 and 4-26.

The same conversion was also applied to the 7.92 mm barrel and bolt of the Model 1924 military rifle, see Figure 4-27. This made it possible to convert the standard military rifle into a training rifle suitable for practice by shooting enthusiasts, soldiers, and young people. On 1 March 1937, Lazar Jovanovich received Patent no.12939 retroactive to 1 May 1935. He then offered his design named "SI-LA Patent L. Jovanovich-Target shooting device, cal.5.4 mm" to the Army and to civilian shooting associations. The Sokol's Association showed immediate interest. In 1938, Colonel Dimitrije Pavlovich, an officer of the Shooting Section of the Sokol's Association, went so far as to incorporate the small-bore, 5.4

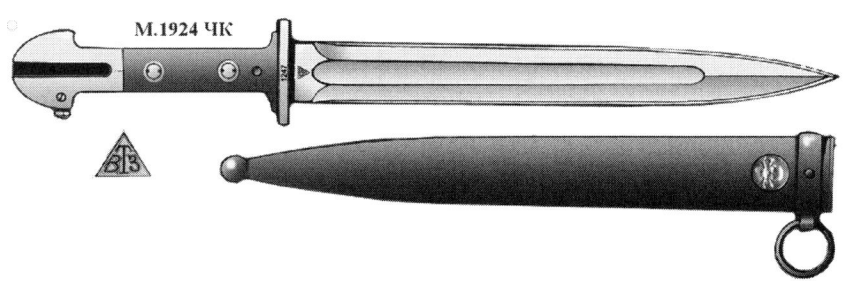

Figure 4-22. Model 1924 *Chetnik* bayonet and scabbard.

Serbian and Yugoslav Mausers

Sokolska puška 7,9 mm   Mauser M1924

Figure 4-23. Model 1924 Sokolska rifle.

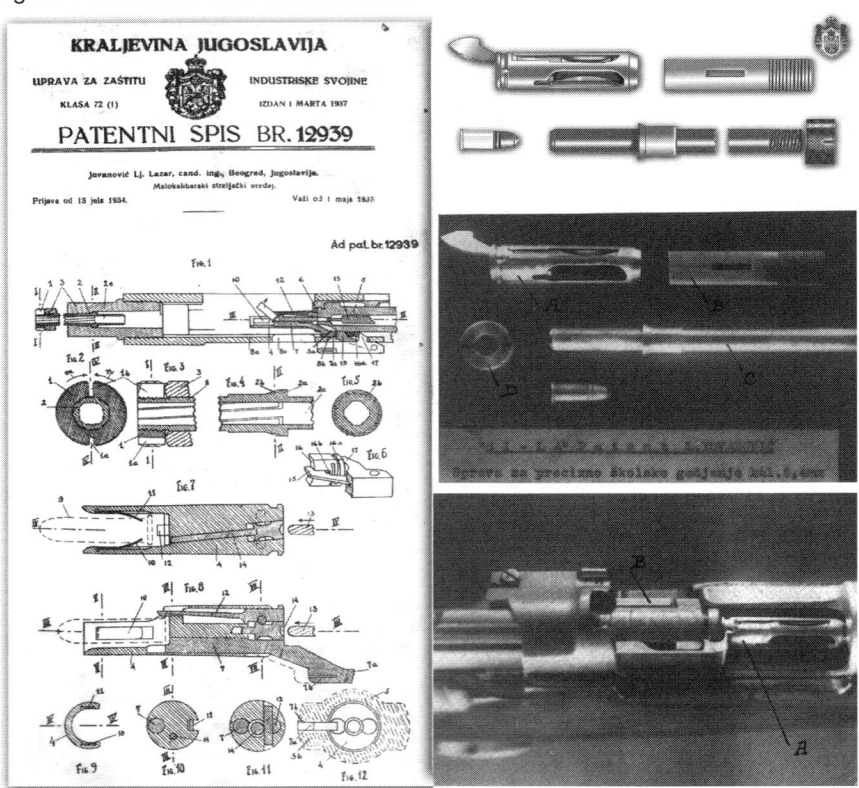

Figures 4-24 through -27. Patent drawings and parts of the 5.56 mm conversion device for the SI-LA system designed by Lazar Jovanovich.

mm rifle into the Sokol training schedule and set about converting 7.92 mm Model 1924s with Jovanovich's device.

The Kingdom's Sokol Association proposed that the Military Technical Institute at Kragujevac mass-produce the devices at an estimated cost of 260 dinars ($4.72) each. But the previous year, the Military

Serbian and Yugoslav Mausers

Figure 4-28. Lazar Jovanovich (center) demonstrates his M.L.J. rifle design at Kragujevac.

Technical Institute had developed and manufactured a complete target rifle designed using the Lazar Jovanovich patents, see Figure 4-28.

The target rifle was marked on the upper part of the receiver ring with the Kingdom's coat of arms and the pattern mark:

Model L.J.

and on the receiver's left side, the manufacturer's name in Cyrillic letters

Voj. Teh. Zavod Kragujevac
(Military Technical Institute-Kragujevac)

Figure 4-29 provides illustrations of the Model L.J. rifles. Unfortunately no information exists as to the number of small-bore devices and target rifles of the Lazar Jovanovich system that Kragujevac manufactured.

Serbian and Yugoslav Mausers

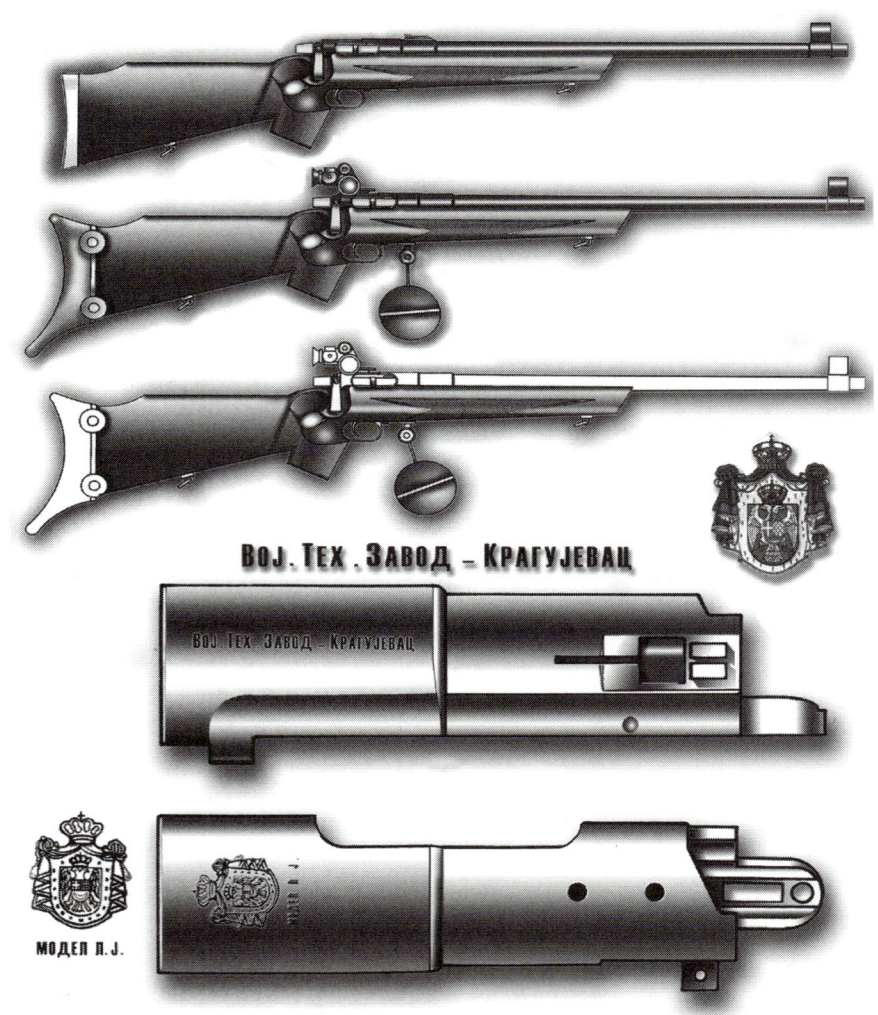

Figure 4-29. Lazar Jovanovich M.L.J. system rifle, manufactured at VTZ Kragujevac, 1931.

CHAPTER 5
MAUSER AND OTHER RIFLE CONVERSIONS

According to a French report dated 15 June 1929, Yugoslavia possessed a plethora of models and calibres of small arms, see Table 5-1.

Table 5-1
Yugoslavian Small Arms Inventory, June 1929*

Model	Number	Calibre	Country of Origin
Mauser Mle. 1924	100,000	7.92 x 57 mm	Belgium
Mauser Model 1924	11,400	7.92 x 57 mm	Yugoslavia
Mauser vz. 24	42,000	7.92 x 57 mm	Czechoslovakia
Lebel Mle.90, 92, 86/93, 07/15, 16, Mle.1874 Gras/ M80. No. 14	143,500	8 mm Lebel	France
Mannlicher 95M, 88/90, 90M	185,000	8 mm Mannlicher	Austria
Mannlicher-Carcano M1891	11,000	6.5 and 7.92 mm	Italy
Mosin-Nagant M1891	34,000	7.62 x 54R mm	Russia
Mauser Model 99C (M99S), Model 99/07, Model 80/07C (Model 80/07S), Model 1910	26,465	7 and 7.92 x 57 mm	Serbia
Mauser Model 1912, 1924B	NA	7 and 7.92 x 57 mm	Mexico
Mauser Model 90/M96T, Model 93/M99T, Model 1903/M9T	NA	7.65 and 7.92 mm	Turkey

* The report did not include an unknown number of single-shot Mauser-Milovanovich M1880 10.15 mm, Werndl Model 1867 and 1873 11 mm and Gras Model 1874 11 mm rifles.

Before the Kragujevac factory focused exclusively on the production of domestically manufactured new Mauser rifles, the March 1923 decision to convert all of Yugoslavia's rifles and carbines to fire the 7.92 x 57 mm cartridge had kept the factory busy to the end of February 1927. It managed to convert 20,732 Serbian Model 99C and M1910C, Mexican 7 x 57 mm M1912/Model 1924B, Turkish M90/M96T and Model 1893/

Serbian and Yugoslav Mausers

M99T, M1903/M9T, and Italian 6.5 x 52 mm Mannlicher-Carcano Model 1891/M1891I rifles, to fire the 7.92 x 57 mm cartridge. Once the factory shifted to domestic production of the Model 1924 Mausers, the government had to find other firms to continue the conversion work.

In order to make these small arms useful as war reserve stocks, on 22 August 1933, the Ministry of the Army and Navy ordered the conversion of all remaining rifles to the Model 1924's performance standards. As of August 1933, the following weapons had yet to be modified: the German Mauser Model 1898 in 7.92 x 57 mm; the French Lebel Model 1886/93F, Model 1907/15F, and Model 1916F rifles, the French Model 1890F musketoon, the French Model 1892F, Model 1907/15F, and Model 1916F carbines (the suffix F denoted French origin); and the Russian 7.62 x 54 mm Mosin-Nagant Model 1891R (the suffix R denoted Russian origin) rifles. Since the Serbian Mauser-Milovanovich-Djurich Model 1880/07 in 7 x 57 mm and the Austrian Mannlicher Model 1888/90 in 8 x 50R mm were obsolete, they were written off as military weapons and issued to the Sokol associations for practice.

The conversion work accomplished in the early 1920s at Kragujevac did include the Austrian Mannlicher 8 x 50R mm Model 1890M and Model 1895M (the suffix M denoted the Mannlicher system) carbines which had been received as war reparations. The Ministry directed that Department V of the Military Technical Institute-Kragujevac adopt a limited number of rifles to serve as models for private contract firms who would then do the actual conversion work. But, as Kragujevac was the only domestic factory capable of working the tough imported steel used in manufacturing Model 1924 barrels, it was decided that they would send the private contractors the finished barrels, stocks, and Model 1924 sight assemblies. See Table 5-2 for technical data concerning the Model 1895M rifle at the end of this section.

THE MAUSER MODEL 1924 B 7.92 MM RIFLE
(ORIGINALLY THE MAUSER 7 MM M1912)

Mexican President Francisco Indalecio Madero ordered a large quantity of modern Mauser Model 1912 rifles from the Austrian firm OEWG in 1912. Newly elected as president to replace Porfirio Díaz, the purchase was intended to strengthen the national army which had been fighting civil war revolutionary forces since 1910.

The German Model 1898 Mauser rifle served as the basis for the Mexican Model 1912 Mauser. The Model 1912 was chambered for the 7 x 57 mm cartridge widely used throughout Latin America, and with an

Serbian and Yugoslav Mausers

updated sight adjustable from 300 to 2,000 meters (328 to 2,187 yards). In all other respects, it was the same as the German Model 1898 rifle.

Markings on the receiver ring included the Mexican coat of arms above the inscription:

REPUBLICA MEXICANA 1912

On the left side of the receiver was the manufacturer's name:

WAFFENFABRIK STEYR AUSTRIA

plus the "S" mark indicating the use of the spire-point bullet. The four-digit serial number was preceded by a Latin letter indicating a run of ten thousand, beginning with "A," see Figure 5-1.

Delivery began in late 1913 or early 1914 but Mexico erupted once again into civil war when General Victoriano Huerta overthrew Francisco Madero's government on 18 February 1913. A general state of lawlessness ensued as factions vied for power. Tensions increased to such an extent along the Mexican-United States border that by early 1914 American President Woodrow Wilson took sides

Figure 5-1. Mexican Model 1912 receiver showing Mexican national crest, Model year, serial number and factory name.

against Huerta and ordered the U.S. Navy to occupy Vera Cruz and impose a naval blockade. Not surprisingly, in this situation delivery of the Model 1912 rifles could not go forward and the Mausers remained at the Steyr factory.

General Huerta abdicated on 8 July 1914 but twenty days later Austro-Hungary declared war on Serbia, and the First World War began. The Austro-Hungarian War Ministry immediately bought the complete consignment of 66,979 Mexican 7 mm Mauser Model 1912s, plus 5,000 Colombian 7 mm Mauser rifles, for 84 cruna (88 dinars or $17.60) each, together with 12,000,000 7 x 57 mm cartridges. These arms, designat-

Serbian and Yugoslav Mausers

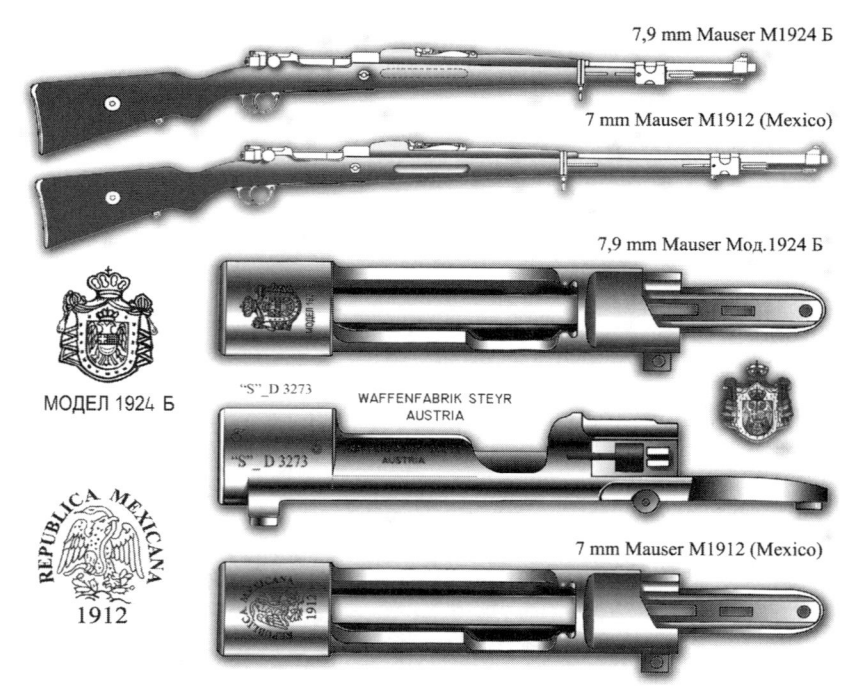

Figure 5-2. Above: Mexican Model 1912 converted to the SHS Model 1924Ъ. Below: an unconverted Mexican Model 1912 rifle. Receiver markings for both rifles are also shown.

ed as the Austrian Model 1914, were supplied to the Austro-Hungarian army. With the collapse of Austro-Hungary in late 1918, all remaining Model 1912/1914 Mausers were transferred to the new nations formed from what had once had been the Austro-Hungarian Empire.

Yugoslav military nomenclature referred to the "Mexican Rifle 7 mm," and included them as part of the rifles designated for conversion to 7.92 x 57 mm. In fact, while these were the first to be sent for conversion, it was not until February 1927 that Kragujevac was finally able to complete the process which included replacing their existing 7 mm barrels, rear sights, and stocks with newly manufactured 7.92 mm calibre barrels, the new 200- to 2,000-meter sights, and new Model 1924 stocks, see Figure 5-2.

The knife bayonet furnished with the original rifles was longer than the standard Mexican Model 1912 bayonet but it could fit the "new" weapon (Type Model 1924B), see Figure 5-3. The adapted rifles were

Serbian and Yugoslav Mausers

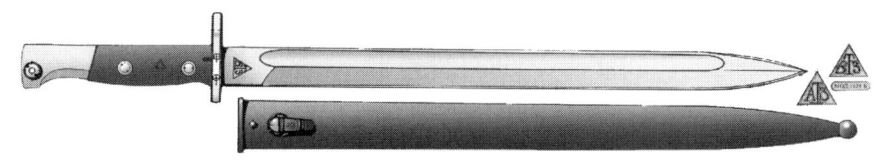

Figure 5-3. Model 1924Б knife bayonet as manufactured at Kragujevac.

designated "МОД1924Б" (Mod. 1924B). When Yugoslavia capitulated early in the Second World War, the German Army seized all Model 24B Mausers and issued them to its forces as the "Gew.291/2[j]."

These rifles were *not* marked on the receiver ring with the Yugoslavian coat of arms and "Model 1912" as has been noted in other publications.

MANNLICHER 7.92 x 57 MM MODEL 1895M, MODEL 1895/24 (ORIGINALLY, 8 x 50R MM MANNLICHER MODEL 1895)

Yugoslavia had received over 185,000 Model 1895 Mannlicher rifles from the former Austro-Hungarian Empire all chambered for the Austrian 8 x 50R mm cartridge, see Figure 5-4. The identifying features of this rifle are shown in Table 5-2 and Figure 5-5. In this calibre, they were of little

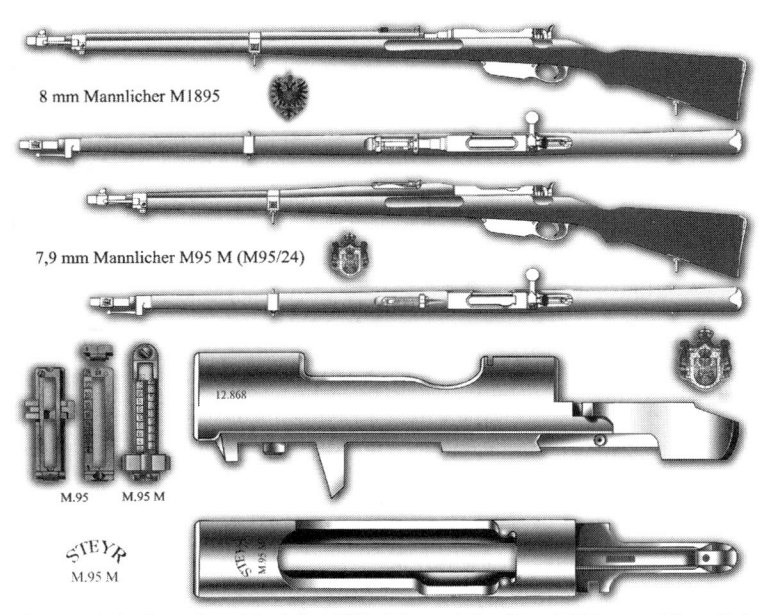

Figure 5-4. Austrian Model 1895 (above) in 8 x 50R mm Mannlicher shown converted to the Model 1895M in 7.92 x 57 mm Mauser (below).

115

Serbian and Yugoslav Mausers

Table 5-2
Markings
Model 1895M 7.9 x 57 mm Mannlicher Rifle

	See Also Table C-1, C-2, C-3, C-4, C-5, C-6, C-7, C-8	Pattern Mark	Manufacturer's Markings	Country's Name	Year of Production	Ruler's Monogram	Full Serial Number	Last Three or Four Digits of Serial Number	Smokeless Powder Proof Mark	Inspection Mark	Acceptance Mark
Barrel							x		x	x	x
Receiver Ring:											
Top		x	x*								
Left							x		x	x	
Right											
Receiver: Austrian Markings											
Bolt Assembly: Austrian marks, except											
Bolt Handle Neck							x				
Bolt Handle Ball									x		
Trigger Assembly: Austrian marks											
Trigger Guard Plate							x			x	
Magazine Floor Plate											x
Stock			x**				x			x	
Butt Plate, Front Band, Lower Band: Austrian marks											

X* Austro-Hungarian: Steyr (OWEG) or Budapest (Fegyvergyár)
X** Yugoslavian: Military Technical Institute, Kragujevav (BT3=VTZ)

Serbian and Yugoslav Mausers

Figure 5-5. Austrian Model 1895/SHS Model 1895M: points of identification

use to the Kingdom of Serbs, Croats and Slovenes/Yugoslavia which had standardized on the 7.92 x 57 mm cartridge. Converting these rifles was beyond the capacity of the Military Technical Institute at Kragujevac, and so the project was put out for bids to commercial companies. Of those competing for the conversion contract in 1933, only the Yakov Poshinger Arms and Ammunition Factory (FOMU) in the city of Uzice met the government's technical and financial requirements.

Serbian and Yugoslav Mausers

The first agreement signed in March 1938 entailed the conversion of 10,000 8 x 50R mm Mannlicher Model 1895 rifles and Model 1890 carbines. To finance the agreement Uzice's District Court mortgaged the factory for 600,000 dinars ($10,909) in favor of the Ministry of the Army and Navy. The factory's daily output was 50 rifles and they were inspected and passed by specialists from the Military Technical Institute.

The rifles were completely disassembled and the Austrian sight assemblies removed and replaced with new 7.92 mm Model 1924 barrels and rear sights shipped from Kragujevac. Other alterations were in keeping with the switch to 7.92 mm calibre and the Model 1924sS cartridge, which imposed size and ballistic considerations somewhat different from the Mannlicher 8 x 50R mm Model 1893 cartridge.

Safety concerns led to drilling two gas relief holes at the rear of the narrower part of the bolt head to permit the gas to escape if the cartridge case split. The use of gas ports to allow gases from split cartridge cases to escape safely was patented by Mauser on 18 August 1895 (DRGM 54,786 & DRGM 56,068). The Austrian Mannlicher bolts did not have gas ports and so workers at FOMU reamed out the lubrication holes in the original Mannlicher bolts.

One problem with the new barrel involved the differences in length between the Austrian 8 x 50R mm cartridge and the Yugoslavian 7.92 x 57 mm cartridge. The Yugoslav cartridge was 2.2 mm (0.087 inch) longer, see Figure 5-6. The Model 1924 cartridge protruded from the

Figure 5-6. Left, Austrian Model 1893 8 x 50R mm cartridge; right, Yugoslav Model 1924 7.92 x 57 mm cartridge.

chamber 3.1 mm (0.122 inch) whereas the original Model 1893 8 x 50R mm cartridge protruded only 1.3 mm (0.05 inch). The extra length of the Yugoslav cartridge exposed the base of the cartridge as far forward as the extractor ring. In addition, the Model 1924 cartridge's base was

Serbian and Yugoslav Mausers

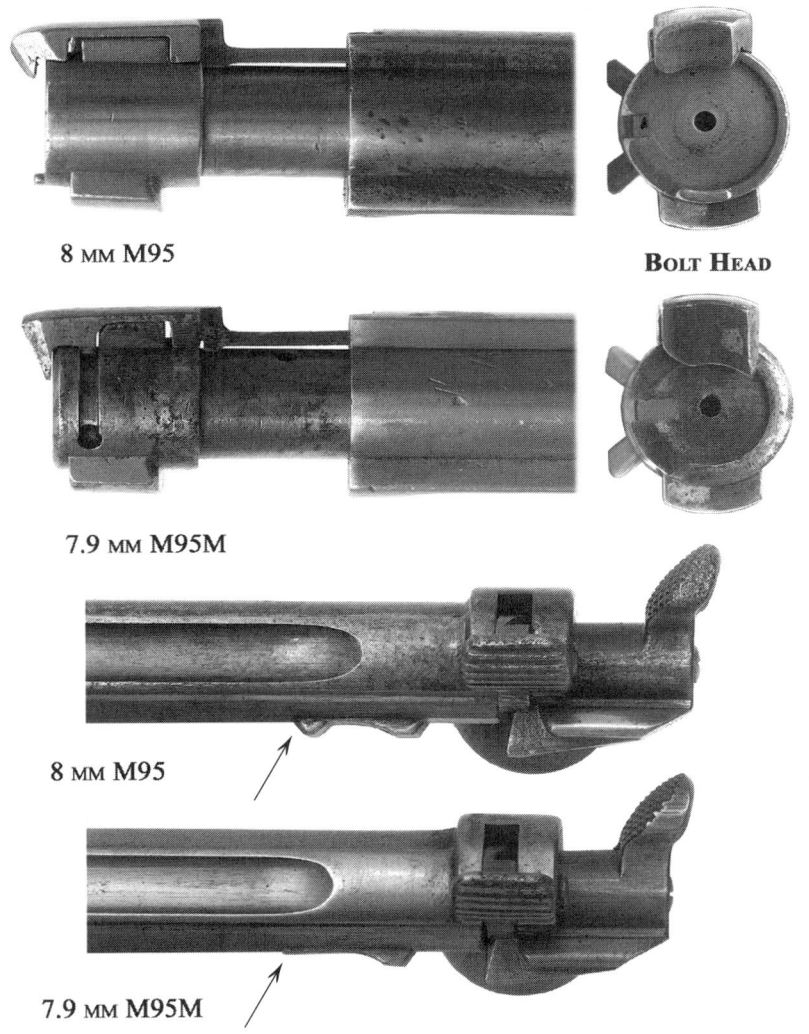

8 mm M95

Bolt Head

7.9 mm M95M

8 mm M95

7.9 mm M95M

Figure 5-7. Austrian Model 1895 and Yugoslav Model 95M bolt cocking pieces with extractors, and bodies compared.

11.9 mm (0.468 inch) in diameter, smaller in diameter than the Austrian Model 1893 cartridge which was 14.1 mm (0.555 inch) in diameter.

The difference in cartridge dimensions forced the design of a new bolt head shorter by 1.8 mm (0.071 inch) than the original, see Figure 5-7.

Serbian and Yugoslav Mausers

The bolt head diameter was also reduced by 2.2 mm (0.086 inch) and the ejector groove deepened and fitted with a new, shorter extractor.

The Austrian Mannlicher system made use of a charger that enclosed five cartridges to load the magazine, see Figure 5-8. To load the rifle, the bolt was pulled back and the charger was pushed down through the open breech into the magazine. The charger remained in the magazine, and when the last cartridge was fired the charger dropped out through the opening in the magazine floor, see Figure 5-9.

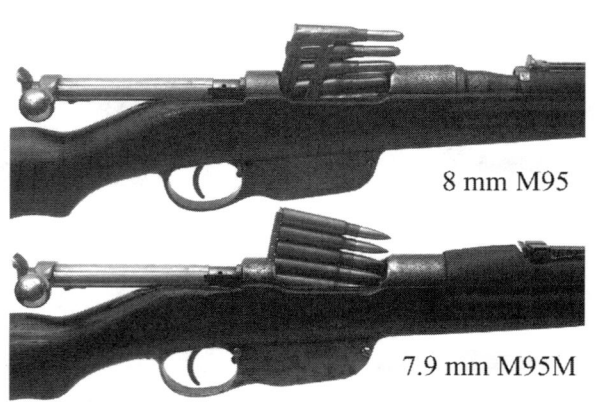

8 mm M95

7.9 mm M95M

Figure 5-8. Above: Austrian Model 1895 Mannlicher-type charger; below: SHS Model 1895M Mauser-type clip.

The Mannlicher magazine system was nearly fifty years old by the time the conversion program was begun and was no longer considered suitable for front-line military use. The straight line design of the charger stacked the cartridges one above the other and forced the magazine box to protrude below the stock. Cartridges could not be loaded individually to "top up" the magazine, nor could the shooter reload until all the rounds in a charger had been fired. Finally, the opening in the magazine floor allowed dirt to penetrate into the receiver.

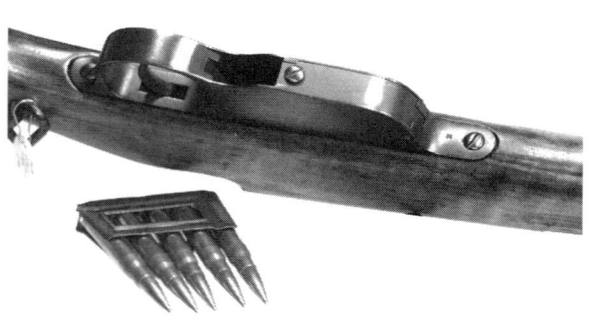

Figure 5-9. Opening in the bottom of the Austrian Model 1895 magazine through which the charger dropped when all cartridges were expended.

Some attempts to correct these problems were made. One solution fixed the Mannlicher Model 1890 charger permanently in the magazine. In some instances where the war reparations rifles lacked their original

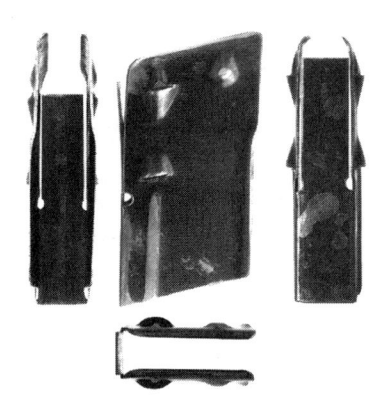

Serbian and Yugoslav Mausers

Figure 5-10. Charger manufactured by FOMU for the Model 1895M. The charger was fixed permanently in the magazine.

magazines, the FOMU factory manufactured a simple sheet metal replacement, see Figure 5-10. They also cut off the useless magazine catch (see Figure 5-11, arrows) and closed the bottom of the magazine with a precisely fitted welded metal cover, bottom photograph in Figure 5-12.

Figure 5-13 shows a comparison cross section of the original Austrian Model 1895 action and magazine assembly and the conversion to the Yugoslav Model 1895M. Notice how the Austrian Model 1890 charger, or its replacement manufactured by FOMU, was fixed permanently (arrow 1) in the magazine. Note also that the magazine catch lost its notch (arrow 2) which secured the charger in place; also that the bottom of the magazine was now closed.

Another modification was made to the left wall of the receiver in the magazine area to permit the rifle to be loaded from a clip rather than a charger, see Figure 5-14. The bottom end of the clip containing five cartridges was inserted into the clip guide machined into the receiver bridge. The shooter pushed the cartridges down into the Model 1890 charger which was now

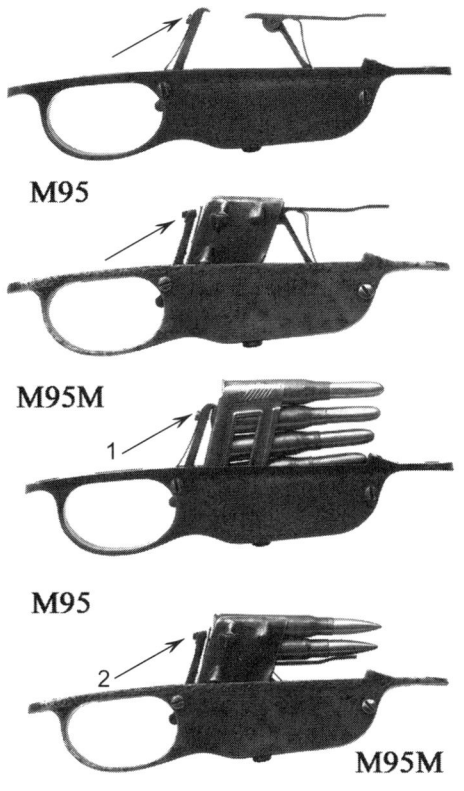

Figure 5-11. Changes to the Austrian Model 1895 magazine assembly to use the SHS five-round clip. Note that the magazine catch was removed (arrows) from the M95M.

121

Serbian and Yugoslav Mausers

permanently fastened into the magazine. This technique had been developed in Germany during the modernization of the Model 1888 "Commission" rifle into the Model 1888/05 and the Model 1888/14. It was also subsequently applied to all Mausers manufactured after 1893. The clip remained in the magazine guide until the last cartridge was fired and then it dropped out when the bolt closed again.

Figure 5-12. A machined floor plate was attached to the SHS Model 1895M magazine to cover the charger ejection opening.

Converted Mannlicher rifles also received a new wooden handguard that was shaped to the new barrels' dimensions and had an opening cut for the rear sight, see Figure 5-15. Uzice finished the conversion project in mid-1939.

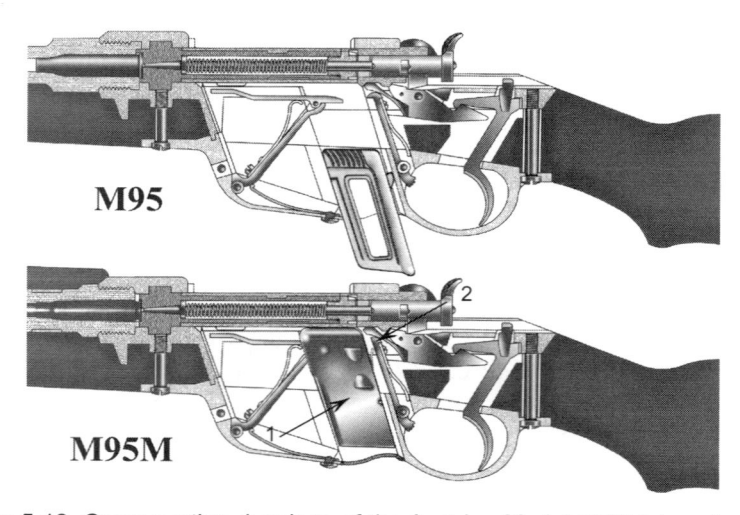

Figure 5-13. Cross section drawings of the Austrian Model 1895 (above) and the SHS Model 1895M (below) showing magazine and charger/clip operation.

Serbian and Yugoslav Mausers

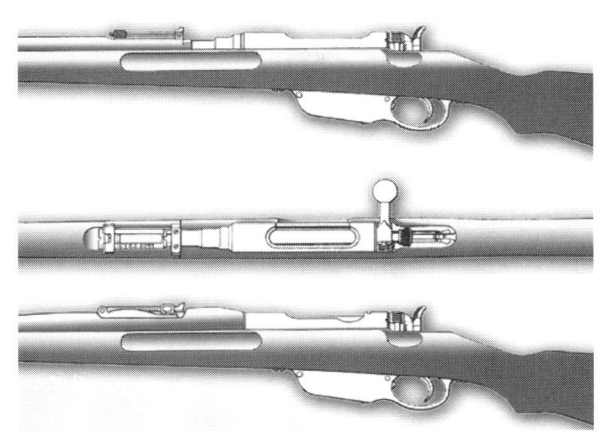

Figure 5-14. The left receiver wall of the Austrian Model 1895 (above) was modified with a thumbcut (below) to make it easier to load a clip of cartridges.

The converted Mannlicher rifles carried a variety of markings, refer to Figures 5-4 and 5-5. The receiver ring retained the original manufacturer's markings, "Steyr" if manufactured at OEWG, or "Budapest" for rifles and carbines made at Fegyvergyár factory. All were marked with the new model marking "M.95M," below the factory name.

The Yugoslavian Army's Ordnance Department, however, did not permit the Uzice factory to add its own mark to the rifles it converted. See Table 5-2 for technical specifications of the Model 1895M rifle.

Regardless of other modifications, all converted Mannlicher rifles retained their original barrel bands, including the distinctive Mannlicher upper band with its machined, projecting straight arm for stacking arms.

The Military Technical Institute at Kragujevac manufactured a very faithful copy of the Model 1895 bayonet for the altered Mannlicher rifles. The only difference between the original and the Kragujevac copy was the factory's "VTZ" mark on the ricasso. Later, Kragujevac perfected manufacture of the Model 1924/95 bayonet and it entered production. These bayonets were quite similar to the Austrian-made bayonet for the Mauser rifles

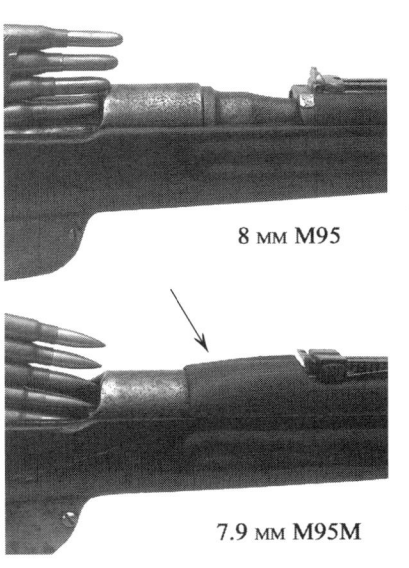

8 мм M95

7.9 мм M95M

Figure 5-15. A new handguard that covered the breech end of the barrel (arrow) was manufactured for the SHS Model 1895M.

Serbian and Yugoslav Mausers

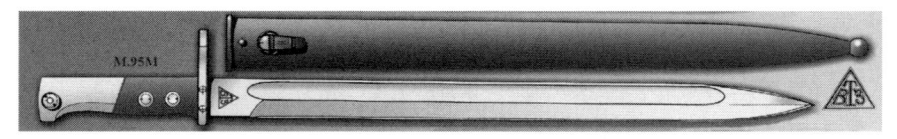

Figure 5-16. Model 1895M knife bayonet manufactured at Kragujevac. Note the "VTZ" marking on the ricasso.

except that their hilt was 22 mm (0.866 inch) shorter, see Figure 5-16. They were issued to the Army with Yugoslav M24 leather belt frogs, see Figure 5-17.

NOTE: The FOMU company did not have the capacity to manufacture bayonets, so the burden of production fell upon the Military Technical Institute at Kragujevac.

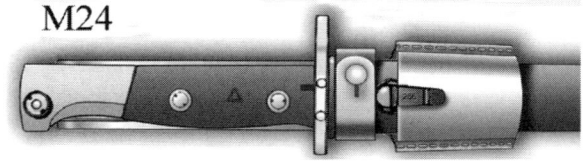

Figure 5-17. Model 1924 belt frog (leather).

Ammunition was carried in stripper clips holding five cartridges. To load the rifle, the bolt was drawn back and the end of the stripper clip

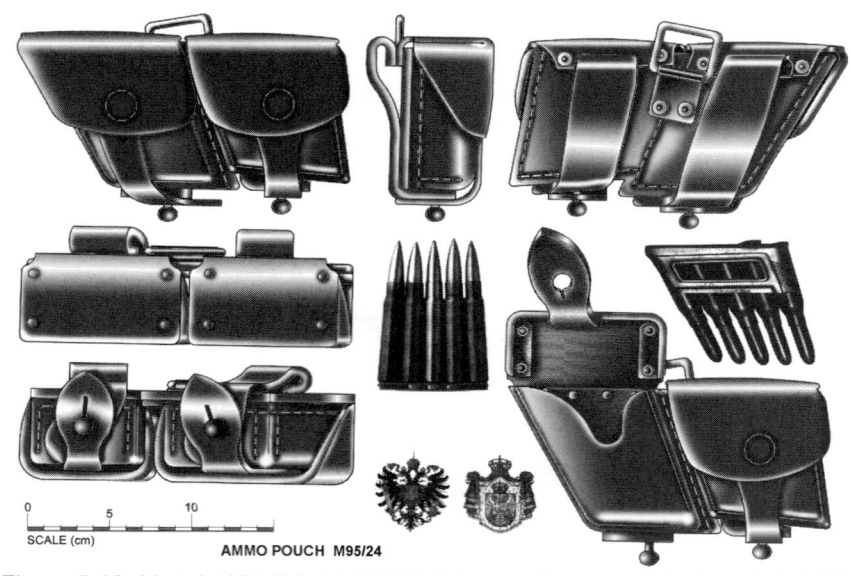

Figure 5-18. Model 1895 (Model 1895/24) Ammunition pouch for the Model 1924 7.92 x 57 cartridge.

Serbian and Yugoslav Mausers

inserted into the cartridge guide in the receiver. The cartridges were then pushed down into the magazine with the thumb.

Ammunition in clips was carried in the Model 1895 ammunition pouch as modified for the Model 1924 equipment. The leather ammunition pouch had twin pockets and was worn in pairs on the soldiers' belts. The Model 1895/24 ammo pouch held a total of 20 cartridges in 4 clips of either the Mauser or Mannlicher ammunition. The boxes were 40 mm (1.6 inches) wide by 170 mm (6.7 inches) long, see Figure 5-18.

The conversion program for Mannlicher rifles received as war reparations produced a very serviceable rifle that was used for training to arm rear echelon and support troops, see Figure 5-19.

In 1941, shortly before Yugoslavia was invaded by Nazi troops, an Austrian Mannlicher Model 1895 8 mm rifle was experimentally adapted at one FOMU plant, to semiautomatic fire, see Figure 5-20. But the outbreak of war made production impossible.

Figure 5-19. SHS Border Guard platoon equipped with Model 1895M Mannlicher rifles.

MAUSER 7.92 MM MODEL 1924C AND MODEL 1924A (ORIGINALLY THE 7.92 MM MAUSER VZ.24)

Between the two world wars, small-arms manufacturing in the Kingdom of Serbs, Croats and Slovenes (SHS)/Yugoslavia paralleled developments taking place in the Republic of Czechoslovakia (RCS). Both countries wanted to force Hungary to give up more of its territory to prevent the

Serbian and Yugoslav Mausers

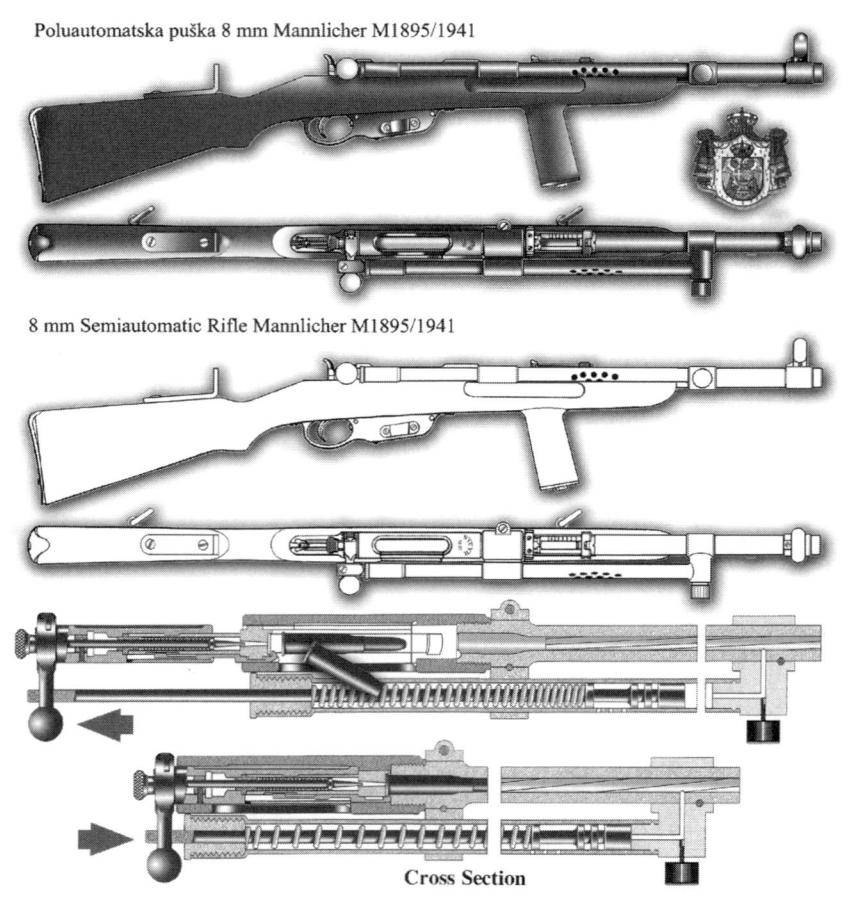

Poluautomatska puška 8 mm Mannlicher M1895/1941

8 mm Semiautomatic Rifle Mannlicher M1895/1941

Cross Section

Figure 5-20. Prototype of the 8 mm Mannlicher M95/41 Semiautomatic Rifle, manufactured by FOMU, Uzice, 1941.

Habsburgs from reestablishing another Austro-Hungarian empire. To accomplish this, Czechoslovakia and the SHS, in early 1919, put together a joint defensive plan. Their delegations quickly decided to equip their armies with the same small arms, all of which would be manufactured in Czechoslovakia. They signed an alliance known as the "Little Entente" on 14 August 1920 and negotiations for sharing arms technology intensified. Subsequently, both countries committed themselves to "thoroughly inspect a rifle design by First Lieutenant Yelen that was just being considered as a possible weapon for the Czech army."

Serbian and Yugoslav Mausers

The Yelen Design
First Lieutenant Rudolf Yelen of the Czechoslovakian Army offered his design to his Ministry of National Defense in August 1919. He based his rifle on the standard German Mauser Model 1898 but reduced it to carbine length and changed the sights to a rear sight having a semicircular notch and an oblique foresight. Lieutenant Yelen selected a turned-down bolt handle and a hinged magazine floor plate and catch assembly. The most distinctive detail of the Czech design, however, was a nose cap that resembled that found on the British Lee-Enfield. He also equipped with his rifle with a front sight hood, a bayonet lug, and a short bar at the tip of the forearm for stacking rifles.

Lieutenant Yelen registered his design at the patent office, and the Brno factory manufactured an experimental model and turned it over to the Army for testing. Initial results failed to impress the Ministry of National Defense's Military-Technical Board. The board concluded that the weapon was little more then "a shortened version of the Mauser rifle with details either taken from the foreign patterns or invented by Lieutenant Yelen," and as such, contained no real innovation that would justify reequipping the Czech Army.

Although there was some truth to their findings, the Military-Technical Board's cold reception also partly reflected the Czech Army's financial situation. Like the SHS, the Czech military had inherited a large quantity of Mannlicher rifles as war reparations. The new nation did not have the financial resources at the time for the Army to do anything other than accept what it believed was a poor Austrian weapon. When a better weapon came along, military leaders were quick to adopt it.

Czechoslovakia received 5,000 of the Steyr-manufactured Mexican Mauser Model 1912 rifles in 7 x 57 mm calibre mentioned at the start of this chapter, from Austro-Hungary. Then, at the end of 1920, Czechoslovakia purchased 77,000 original 7.92 x 57 mm German-made Mauser Model 1898 rifles from France. These are sometimes referred to as the "Bavarian rifles." Another supply of 1,500 Mausers soon arrived from the Oberndorf factory in the form of war reparations but, on that occasion, the Czechs also bought up ready-made parts sufficient for assembling 42,000 more Model 1898 7.92 x 57 mm rifles.

Another large supply of Mausers was soon available. Just prior to the armistice in November 1918, several German divisions withdrew to neutral Holland and surrendered their weapons to the Dutch rather than allow them to be captured by the Allied forces. Holland also benefited when the German arms giant Rheinmetall Borsig AG decided to disassemble most of its machinery, installations, tools, and rifles, and ship

Serbian and Yugoslav Mausers

everything under false documents to that country shortly before the allied controllers arrived at the factory. This contingent of rifles and equipment was consigned to the Rotterdam bank for disposition and put into storage in the Dutch cities of Rotterdam and Delfzijl. Control over this property was transferred from the Rotterdam bank during 1922 and 1923 to the Schweizer Waffenfabrik Solothurn factory in Switzerland. Prior to this, in June 1921, agents from the firm of Bernhardt & Steinhardt (IHAG) offered to sell some of the Mausers held by Holland to the Czechoslovakian government. A year later, an agreement was reached and Czechoslovakia acquired 60,000 more Model 1898 7.92 x 57 mm Mauser rifles. The weapons were quickly transported to its arsenal at Pirkartice.

Czechoslovakia Selects the Model 1898 Mauser Action
As the number of Mausers in Czech arsenals increased, the Ministry of Defense decided in April of 1920 to equip its entire Army with Mauser rifles. As a consequence, Ministry officials took another look at Lieutenant Yelen's design and ordered three of his rifles from the Brno factory in 7 x 57 mm calibre. The three weapons were not in fact the weapon designed by Yelen, but were instead Mexican Mauser Model 1912s that had been slightly altered with some of Yelen's design. Testing of the three modified Model 1912s proved encouraging and, on 8 September 1920, a Ministry of Defense commission ordered 150 more modified Mexican Model 1912s. As a consequence, a joint Czech-SHS commission decided to retest Yelen's original rifle and, in February 1921, the Brno factory received an order for 150 Mauser-Yelen rifles in 7 x 57 mm calibre and another 150 in 7.92 x 57 mm calibre.

Brno began work on the order in October 1921 and finished at the end of January 1922. Regardless, an extremely cold winter delayed full testing at the Milovice rifle range for some time and not until April 1922 did the Kingdom of Serbs, Croats, and Slovenes (SHS) receive its 150 "Yugoslav" rifles. After further testing, the Czechoslovakian commission turned in an unfavorable report to the Ministry on 30 September 1922. It found the rifles to be expensive and unsuitable for mass production while offering no real improvement over the original Mauser weapon. With this report, the Czechoslovakian government ceased further testing.

While nothing is known about the final disposition of the 150 Yelen rifles sent to Belgrade, the SHS government also abandoned Yelen's design and, in the spring of 1923, ordered 50,000 of the Model 1898 7.92 x 57 mm Mausers that Czechoslovakia had acquired earlier from Bavaria.

The SHS commission's decision was probably influenced in great part by the offer in 1922 by the Czech Skoda factory's offer of the Mauser

Serbian and Yugoslav Mausers

production machinery it had received from Germany as part of war repatriations.

Even though Czechoslovakia had expended some effort in testing Yelen's rifle, it was making a greater effort to establish the domestic capability for manufacturing the Model 1898 7.92 x 57 mm Mauser. The machinery from Oberndorf enabled the Czechs to begin production as early as the autumn of 1920. But it was not until January 1922 that the Brno factory received an order for 1,000 "long" Mausers from the Ministry of Defense. Almost immediately that order was increased to 40,000 rifles even though the commission had not yet settled on the rifle's final design.

The Czechoslovakian vz.23 Mauser Rifle

That decision came a month later. The new rifle was to be the Model 1898 Mauser, as expected, but equipped with the "Mexican" Mauser's Model 1912 rear sight (the so-called "Kurven-Schiebeviser" sight) graduated from 300 to 2,000 meters (328 to 2,187 yards). It took some time for that decision to become official and by then Brno had produced 10,000 rifles with the original Model 1898 sights, sometimes referred to as the "Lange Vizier" sight that was graduated from 200 to 2,000 meters (219 to 2,187 yards). The "Mexican sights" were mounted on the remaining 30,000 rifles, which, to avoid confusion, were designated as vz.98/22 (vz. for *vzor* or model).

Debate nevertheless continued over the exact performance standards desired. The commission finally reached a decision that resulted in an announcement on 2 October 1922 from the SHS Chief of the General Staff to the effect that the shortened Mauser vz.98 rifle would be the Army's standard weapon. The rifle with attached bayonet could not exceed 1,500 mm (59.05 inches) while the original Model 1898 was 1,750 mm (68.9 inches) long. Also, the bullet fired from that weapon had to pierce a steel artillery-gun shield or machine-gun shield at a distance of 500 meters (547 yards).

The new rifle's calibre was to be 7.92 x 57 mm and it would incorporate the Mexican Model 1912 Kurven-Schiebeviser rear sight. On 30 December 1922, the Army ordered 90,000 "shortened Mausers vz.98" from the Brno factory. Production was interrupted by additional alterations and delivery was delayed to July 1923. Initially listed as "short rifles vz.98" they were redesignated as the "vz.98/23," and finally as the "vz.23."

Serbian and Yugoslav Mausers

Problems continued. Even though the Czech Army had decided on the Mauser vz.23 as its pattern, it delayed work on the 7.92 x 57 cartridge until the second half of 1923. Tests quickly showed that the Model 1912 sights mounted on the 80,000 new rifles that had been received were incompatible with the ammunition's performance characteristics. The arsenal at Brno responded quickly and developed a new sight calibrated for the new ammunition's ballistics. But this rear sight was only mounted on the 10,000 vz.23 rifles remaining to be produced while the previous 80,000 vz.23s retained their Model 1912 sights. To avoid confusion in the field, the final 10,000 weapons were designated "vz.23a."

The Czechoslovakian vz.24 Mauser Rifle

The Brno factory also reconciled the sometimes-different needs of the cavalry and infantry in carrying the rifle by mounting an additional sling swivel on the left side of the buttstock, an idea copied from the Mannlicher Model 1895 carbine. Thus modified, the rifles made after September 1924 were designated by the Czechs as the "vz.24." The Army had placed its first order for 40,000 of what became the vz.24 rifles on 30 December 1923 and the second order for 50,000 rifles on 31 December 1924. The Kingdom of Serbs, Croats, and Slovenes (SHS) would benefit from Czechoslovakia's third order for 15,000 and its fourth order for 45,000 vz.24 rifles made in October and November 1925.

By 1925 relations between the SHS and fascist Italy had worsened to the point of crisis. That was not a favorable moment for the SHS Army, as the Kragujevac factory had just received the machinery needed for its Mauser production and the 100,000 rifles purchased from Belgium were insufficient to rearm the entire Army with modern weapons. As Czechoslovakia had begun domestic production of the vz.24 rifle, the Belgrade government turned to Prague for support. In March 1925, King Alexander I, along with Dr. Momcilo Ninchich, the Minister of Foreign Affairs, and General Dusan Trifunovich, the Minister of War of the SHS, asked Yan Sheba, the Czech Ambassador, for his country's immediate support with 100,000 new rifles and 100,000,000 cartridges. Although sympathetic, Czechoslovakia could not satisfy this request as it was also in the process of rearming its own army. In November 1925, the SHS government reduced its request to between 50,000 and 60,000 rifles and an agreement was reached between the two countries in late January 1926.

The Czechoslovakian Ministry of Defense agreed to defer rearming its army if the SHS government immediately placed its order with the

Serbian and Yugoslav Mausers

Brno factory. Czechoslovakia pledged to deliver their entire third order of 15,000 vz.24s as well as 27,000 rifles from the fourth order which had completed manufacture in November 1925. They also offered to transfer 20,000 of the old Model 1898 Mausers which were in storage at Pirkartice. The Brno factory committed itself to the Czech army to produce an additional 15,000 new vz.24 rifles to compensate for the lost 20,000 Model 1898s. By mid 1926, Czechoslovakia had delivered some 42,000 modern vz.24 rifles to the SHS as well as 20,000 of the Great War-vintage Model 1898 Mausers.

On the basis of a loan obtained in 1928, the SHS committed itself to purchase more than 50,000 new vz.24 rifles from Czechoslovakia on 10 June 1929, the same year that the name of the country was changed from the Kingdom of Serbs, Croats and Slovenes (SHS) to Yugoslavia.

Initially, the Yugoslav nomenclature for the ultimate total of 92,000 vz.24 Czechoslovakian rifles, also called *puske* rifles, 7.92 mm calibre, was

Model 1924Ч

The Cyrillic letter "Ч" is pronounced as "Ch" and indicated Czechoslovakian manufacture. After 1927, however, the weapon came to be officially referred to as

Carbine, 7.92 mm, Model 1924a
(or sometimes Model 1924A)

The vz.24 sling swivel position on the side of the buttstock corresponded to that found on the domestically manufactured carbine Model 1924; consequently, the Yugoslav army classified the Czech-made weapons as carbines. The vz.24 became the principal weapon issued to Yugoslavian gendarmes, see Figure 5-21. The characteristics of the Czech rifle were almost identical to the domestically manufactured Model 1924 with the most notable difference being the rear sight, which was graduated from 300 to 2,000 meters (328 to 2,187 yards). The receiver ring of the Czechoslovakian rifles was marked

CESKOSLOVENSKA/ZBROJOVKA/=BRNO=

with the model mark "vz.24" stamped on the receiver's left side.

Serbian and Yugoslav Mausers

Polish Mausers

Not much is known about the Mausers acquired from Poland, neither when purchased nor in what quantity. It is likely that most did not enter Yugoslavia until the 1940s. It is known that in 1932 Yugoslavia and Poland signed a contract on exchange armaments. In return for a large quantity of the Roth-Steyr Model 1907 and Steyr Model 1912 pistols, Poland delivered to Yugoslavia the wz.29 rifles produced in 1934 at Radom.

Figure 5-21. A Yugoslav gendarmerie patrol equipped with their standard-issue Czech Model 24a (vz.24) rifles in 7.92 x 57 mm calibre.

Those examples found in Yugoslavian museums date to 1934 and are stamped with the Polish national eagle, place of manufacture, and date of manufacture on the receiver ring, i.e., "F.B./RADOM/1934." The model marking and serial number were stamped on the left side of the receiver: "wz.29" followed by the serial number.

According to an inventory conducted in 1937 and included in *The Regulation on War Armament of the Kingdom of Yugoslavia,* the Army listed no Polish Mauser rifles in its possession.

Serbian and Yugoslav Mausers

MAUSER 7.92 MM MODEL 1924B
(FORMERLY THE MAUSER M1898 7.92 MM)

At the end of 1939, the Yugoslavian War Ministry directed that the remaining 71,000 German Mauser Model 1898 7.92 x 57 mm rifles would be converted to Model 1924 7.92 x 57 mm. Initially, in 1933, the Ministry had decided that the Army would not convert these German Model 1898 Mausers held in war reserve as they were already capable of firing the domestic 7.92 x 57 mm sS Model 1924 cartridge. Five years later, however, the Technical Department of the War Ministry concluded that the German rifles were too long for practical service and lacked the characteristics of the Mauser carbines adopted for the regular Army.

Figure 5-22. Rifle Model 1924B workshop, FOMU, Uzice.

A commission, formed to decide on either the further production or the purchase of firearms, met on 28 January 1941. It noted that the private factory, FOMU, had received a large contract to modify the Model 1898s as well as manufacture a large quantity of ammunition, see Figure 5-22. For this work, the government allocated a total of 15,420,980.52 dinars ($280,381.46). The Cabinet decided to continue the project, but noted that the project had not been without complications for the private company. When he had secured the original contract previously, the factory's owner, Yakov Poshinger Jr., had to seek additional operating

Serbian and Yugoslav Mausers

capital. His efforts proved successful and towards the end of 1939, Uzice's District Court approved a mortgage of 6,500,000 dinars in favor of the "Belgrade Credit Union, Inc." The need to convert the old Model 1898s to the standards of the Model 1924 was real. The Gew.98 rifle had been chambered originally to fire the 7.92 x 57 mm calibre Model 1888 round nose (ogival-cylindrical) bullet weighing 14.7 grams (226.8 grains). The cartridge was loaded with 2.75 grams (42.4 grains) of Gew. Bl.P.88 smokeless powder which produced at 25 meters (27.3 yards) a velocity (V25) of 620 meters per second (2,034 feet per second). This required a barrel length of 740 mm (29.1 inches) to properly burn the powder. The barrel length then dictated a complete rifle length of 1250 mm (49.2 inches) which made it very unwieldy for the needs of modern warfare.

When the new spire-point "S" cartridge was adopted in 1904, the rifle had been rechambered for the slightly greater diameter bullet. The magazine and rear sight had also been changed but the barrel and overall length remained the same. These then were the rifles that had been received from the Czech government, who had in turn received them from Germany as war reparations.

The technical commission of the Ministry of the Army and Navy of the Kingdom of Yugoslavia in 1939 came to several conclusions regarding what needed to be done to the Model 1898 rifles to bring them up to the standards of the Model 1924. It requested that the "Lange" sight assembly be graduated from 200 to 2000 meters (219 to 2,187 yards) or the later "S" version graduated from 400 to 2000 meters (437 to 2,187 yards) and the long 740 mm (29.1-inch) barrel be replaced by the shorter 589 mm (23.2-inch) Model 1924 barrel with the Model 1924 rear sight.

It also wanted the stacking hook on the nose cap removed and the existing stock recut to the Model 1924's dimensions. The combination of the new Model 1924 barrel and sight assembly required a new handguard. This new handguard extended from the front of the receiver to the lower band position. The original barrel bands were reused. This differed from the domestically manufactured Model 1924 rifles in which the handguard extended from the receiver to the upper band.

Since there was insufficient time to manufacture the bayonets required for the converted rifles, the commission further decided to simply modify the original Model 1898 bayonets to approximate the Model 1924's bayonet dimensions by shortening the blade, removing the quillon, and adding a muzzle ring. These bayonets were then able to use the standard Model 1924 scabbard. All Model 1898 Mauser rifles and bayonets converted to these specifications were re-stamped "M.24B" on the receiver ring, see

Serbian and Yugoslav Mausers

Figure 5-23. The guard, or crosspiece, of the bayonet was also stamped "M.24B." They were issued to the Army with the Yugoslav M24 leather belt frogs, see Figure 5-24.

Figure 5-23. Receiver ring markings of a Gew 1898 German Mauser converted to the Model 1924B. The original German markings were machined off.

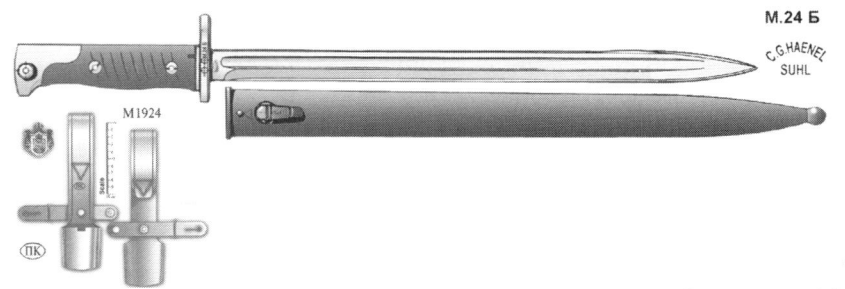

Figure 5-24. Modified German knife bayonet for the Gew 1898 rifle converted for use with the Model 1924B rifle. The original manufacturer's name was left intact and the Yugoslav Model designation and inspector's mark added.

Summary of Mauser Model 1898 Conversions

The exact number of Model 1898 rifles that underwent conversion to Model 1924Bs is unknown. Before the conversion project, Yugoslavia possessed 71,600 Model 1898 rifles. The Uzice factory committed itself to a quota of 1,000 Mauser conversions a month and, by the end of 1939, it had delivered 6,100 Model 1924Bs to the Army.

Assuming that the factory maintained this rate of production, by the time of the German invasion in April 1941, a total of 21,000 converted rifles would have been completed. German officials noted that when their forces seized the factory complex they found an estimated 15,000 to 20,000 dismantled barrels. One inventory indicated a total of 16,050 barrels but, regardless of the exact number, the estimates seem reasonable.

Serbian and Yugoslav Mausers

The Germans did not hold the factory long enough to do much with this material and when the complex came under Partisan control, the "First Partisan's Arms and Ammunition Factory" as it was renamed, could not theoretically have produced more than 15,750 rifles at best, with the more likely estimate being somewhere between 10,000 and 13,000 rifles.

When the German 342nd Infantry Division reoccupied Uzice in December 1941, its staff filed a report stating that there still remained a large quantity of rifle parts ready for final assembly at the factory complex Uzice-East at Krcagovo. The German report "*VI. Fabriken: a) Gewehr-und Munitionsfabrik Uzice ost "Grosse Mengen Halbfabrikate (Waffen . . .)*", suggests that the Partisans failed to use all semi-finished parts and dismantled barrels, which suggests that the original estimate of about 15,000 disassembled rifles in April of 1941 was realistic. Consequently, Uzice's original contract must have been to convert between 35,000 and 36,000 of the old Model 1898 rifles—about 50 percent of the total amount in army arsenals in 1939. Of that 35,000 to 36,000, the factory apparently completed about 60 percent, before Yugoslavia was overrun by the German onslaught.

Mauser rifles and carbines in 7.9 calibre in Yugoslav inventory in 1937 are listed below in Table 5-3. When compared to Table 4-2 compiled in 1929, it is clear that the Yugoslavs had made a great deal of progress in rearming their military forces with compatible weapons of one design and calibre.

Table 5-3 Yugoslavian Army Small-Arms Inventory, 1937				
No	Name	Ammunition	Origin	Note
1	Puska 7mm M10 C (M10 S) Rifle M10S 7mm	7 x 57 mm M14 & M99	Serbian	Serbian rifle 7 mm Mauser M1910
2	Puska 7mm M99/7 C (M99/7S) Rifle M99/7 S 7mm	7 x 57 mm M14 & M99	Serbian	Serbian rifle 7 mm Mauser M99/07
3	Puska 7mm M99 C (M99 S) Rifle M99 S 7mm	7 x 57 mm M14 & M99	Serbian	Serbian rifle 7 mm Mauser M99

Serbian and Yugoslav Mausers

Table 5-3, cont. Yugoslavian Army Small-Arms Inventory, 1937				
No	Name	Ammunition	Origin	Note
4	Puska M80/7 C (M80/7 S) Rifle M80/7 S	7 x 57 mm M14 & M99	Serbian	Serbian rifle 7 mm Mauser-Milovanovich-Djurich M80/07
5	Puska 7mm meksikanska Mexican rifle 7mm	7 x 57 mm M14& M99	Mexican	Mexican rifle 7 mm Mauser M12
6	Puska 7,9mm M24R Rifle M24 7.9mm	7.92 x 57 mm M24/SS, M24/SSB, M24/S	Yugoslav	Yugoslav rifle M24 7.92 mm
7	Puska 7,9mm M98 Rifle M98 7.9mm	7.92 x 57 mm M24/SS, M24/SSB, M24/S	German	German rifle M98 7.92 mm
8	Puska 7,9mm M88 Rifle M88 7.9mm	7.92 x 57 mm M24/SS, M24/SSB, M24/S	German	German "Commission" rifle M88 7.92 mm
9	Puska 7,9mm M95 M Rifle M95 M 7.9mm	7.92 x 57 mm M24/SS, M24/SSB, M24/S	Austrian, Mannlicher System	Yugoslav Mannlicher M95M (M95/24) 7.92 mm- repaired Austrian Mannlicher M95 8 mm
10	Puska 7,9mm M9 T Rifle M9 T 7.9mm	7.92 x 57 mm M24/SS, M24/SSB, M24/S	Turkish	Yugoslav Mauser M9T 7.92 mm- repaired Turkish Mauser M1903 7.65 mm
11	Puska 7,9mm M99 T Rifle M99 T 7.9mm	7.92 x 57 mm M24/SS, M24/SSB, M24/S	Turkish	Yugoslav Mauser rifle M99T 7.92 mm- repaired Turkish Mauser M1893 7.65 mm
12	Puska 7,9mm M96 T Rifle M96 T 7.9mm	7.92 x 57 mm M24/SS, M24/SSB, M24/S	Turkish	Yugoslav Mauser rifle M96T 7.92 mm- repaired Turkish Mauser M1890 7.65 mm

Serbian and Yugoslav Mausers

	Table 5-3, cont. Yugoslavian Army Small-Arms Inventory, 1937			
No	Name	Ammunition	Origin	Note
13	Puska 7,9mm M24 Б Rifle M24B 7.9mm	7.92 x 57 mm M24/SS, M24/SSB, M24/S	Mexican	Yugoslav Mauser rifle M24B- repaired Mexican Mauser M1912 7 mm
14	Puska 7,9mm M10 C (M10 S) Rifle M10S 7.9mm	7.92 x 57 mm M24/SS, M24/SSB, M24/S	Serbian	Yugoslav Mauser rifle M1910S 7.92 mm- repaired Serbian Mauser M1910 7 mm
15	Puska 7,9mm M99 C (M99 S) Rifle M99 S 7.9mm	7 x 57 mm M24/SS, M24/SSB, M24/S	Serbian	Yugoslav rifle M1899S 7.9 mm- repaired Serbian Mauser M1899 7 mm
16	Puska 7,65mm M9 T Rifle M9T 7.65mm	7.65 x 53 mm M90 T, M3 T	Turkish	Turkish Mauser rifle M1903 7.65 mm
17	Puska 7,65mm M99T Rifle M99T 7.65mm	7.65 x 53 mm M90 T M3 T	Turkish	Turkish Mauser rifle M1893 7.65 mm
18	Puska 7,65mm M96T Rifle M96T 7.65mm	7.65 x 53 mm M90 T, M3 T	Turkish	Turkish Mauser M1890 7.65 mm
19	Karabin 7mm M8 C (M8 S) Carbine M8 S 7mm	7 x 57 mm M14 & M99	Serbian	Serbian carbine Mauser M1908 7 mm
20	Karabin 7,9mm M10 C (M10 S) Carbine M10 S 7.9mm	7.92 x 57 mm M24/SS, M24/SSB, M24/S	Serbian	Yugoslav carbine M1910S 7.92 mm- repaired Serbian Mauser M1910 7 mm
21	Karabin 7,9mm M99 C (M99 S) Carbine M99 S 7.9mm	7.92 x 57 mm M24/SS M24/SSB, M24/S	Serbian	Yugoslav carbine M1899S 7.92 mm- repaired Serbian Mauser M1899 7 mm

Serbian and Yugoslav Mausers

	Table 5-3, cont. Yugoslavian Army Small-Arms Inventory, 1937			
No	Name	Ammunition	Origin	Note
22	Karabin 7,9mm M95 M Carbine M95 M 7.9mm	7.92 x 57 mm M24/SS, M24/SSB, M24/S	Austrian, Mannlicher System	Yugoslav carbine M1895M 7.9 mm- repaired Austrian carbine Mannlicher M1895 8 mm
23	Karabin 7,9mm M90 M Carbine M90 M 7.9mm	7.92 x 57 mm M24/SS, M24/SSB, M24/S	Austrian, Mannlicher System	Yugoslav carbine M1890M 7.92 mm- repaired Austrian Mannlicher carbine M1890 8 mm
24	Karabin 7,9mm M98 Carbine M98 7.9mm	7.92 x 57 mm M24/SS, M24/SSB, M24/S	German	German carbine M1898 AZ 7.92 mmS
25	Karabin 7,9mm M24 Carbine M24 7.9mm	7.92 x 57 mm M24/SS, M24/SSB, M24/S	Yugoslav	Yugoslav carbine Mauser M1924 7.92 mm
26	Karabin 7,9mm M24Ч M24C, M24.a Carbine M24C, M24.a 7.9mm	7.92 x 57 mm M24/SS, M24/SSB, M24/S	Czech	Yugoslav carbine M24C 7.9 mm-Czech Mauser rifle vz.24 7.92 mm
27	Karabin 7,9 mm M91И Carbine M91I 7.9mm	7.9x57 mm M24/SS, M24/SSB, M24/S	Italian	Yugoslav carbine M1891 7.9 mm- repaired Italian rifle Mannlicher-Carcano M1891 6.5 mm

CHAPTER 6
OCCUPATION

Germany attacked Yugoslavia on 6 April 1941 without a declaration of war. The Yugoslav army was taken by surprise and resistance was short-lived. The High Command capitulated at 9:00 PM, 17 April. As the Occupation of Yugoslavia began, the Wehrmacht seized a large quantity of the Mauser-system rifles. Kragujevac had 250,000 completed Model 1924 rifles in storage. Using German Mauser nomenclature, the total inventory of arms seized was reported as follows:

		Table 6-1 German Army Inventory of Serbian Small Arms 1941			
No. D.50/1 01(j)	German Name	Yugoslav Name	Calibre	Origin and Model	
221	Gewehr 221(j)	Puska M10C	7 x 57 mm	Serbian Mauser rifle M1910 7 mm	
222	Gewehr 222(j)	Puska M99	7 x 57 mm	Serbian Mauser rifle M1899 7 mm	
223	Gewehr 223(j)	Puska M80/07C	7 x 57 mm	Serbian Mauser Milovanovich-Djurich rifle M1880/1907 7 mm	
288	Gewehr 288(j)	Sokol-Puska 7,9 mm	7.92 x 57 mm	Yugoslav Sokol's rifle M24 7.9 mm	
289	Gewehr 289(j)	Chetniks (Komiten)-Puska 7,9 mm	7.92 x 57 mm	Yugoslav Chetnik's carbine Mauser M24CK 7.9 mm	
290	Gewehr 290(j)	Puska Brno VZ 24	7.92 x 57 mm	Czech rifle Mauser vz.24 7.9 mm	

Serbian and Yugoslav Mausers

No. D.50/1 01(j)	German Name	Yugoslav Name	Calibre	Origin and Model
		Table 6-1, cont. German Army Inventory of Serbian Small Arms 1941		
291/1	Gewehr 291/1(j)	Puska M24	7.92 x 57 mm	Yugoslav Mauser M1924 7.9 mm
291/2	Gewehr 291/2(j)	Puska M24B	7.92 x 57 mm	Yugoslav Mauser M24B 7.9 mm-repaired Mexican rifle M1912 7 mm
291/3	Gewehr 291/3(j)	Puska M10C	7.92 x 57 mm	Yugoslav carbine Mauser M1910S-repaired Serbian Mauser rifle M1910 7 mm
291/4	Gewehr 291/4(j)	Puska M99C	7.92 x 57 mm	Yugoslav carbine Mauser M1899S 7.9 mm-repaired Serbian rifle Mauser M1899 7 mm
292	Gewehr 292 (j)	Puska M88	7.92 x 57 mm	German "Commission" rifle M88 7.9 mm
293	Gewehr 293(j)	Puska M98	7.92 x 57 mm	Yugoslav Mauser M24B 7.9 mm- repaired German M1898 7.9 mm
294	Gewehr 294(j)	Puska 7,9 mm M95M	7.92 x 57 mm	Yugoslav rifle M95M 7.9 mm- repaired Austrian Mannlicher M95 8 mm
295	Gewehr 295(j)	Puska M9T	7.92 x 57 mm	Yugoslav Mauser rifle M9T 7.9 mm- repaired Turkish Mauser rifle M1903 7.65 mm

Serbian and Yugoslav Mausers

	Table 6-1, cont. German Army Inventory of Serbian Small Arms 1941			
No. D.50/1 01(j)	German Name	Yugoslav Name	Calibre	Origin and Model
296	Gewehr 296(j)	Puska M99T	7.92 x 57 mm	Yugoslav Mauser rifle M99T 7.9 mm- repaired Turkish Mauser M1893 7.65 mm
297	Gewehr 297(j)	Puska M96T	7.92 x 57 mm	Yugoslav Mauser rifle M96T 7.9 mm-repaired Turkish Mauser M1890 7.65 mm
298	Gewehr 298(j)	Puska M29	7.92 x 57 mm	Polish rifle Mauser wz.29 7.9 mm
352	Gewehr 352(j)	Puska M78/80	10.15 x 63 mm R	Serbian Mauser-Milovanovich rifle "Kokinka" M1880 10.15
421	Karabiner 421(j)	Karabin M8C	7 x 57 mm	Serbian Mauser carbine M1908 7 mm
491/1	Karabiner 491/1(j)	Karabin M24	7.92 x 57 mm	Yugoslav Mauser carbine M24 7.9 mm
491/2	Karabiner 491/2(j)	Karabin M24 B	7.92 x 57 mm	Yugoslav carbine M24 7.9 mm repaired Mexican Mauser M1912 7 mm
491/3	Karabiner 491/3(j)	Karabin M10 C	7.92 x 57 mm	Yugoslav Mauser carbine M10S 7.9 mm repaired Serbian Mauser rifle M1910 7 mm

Serbian and Yugoslav Mausers

Table 6-1, cont. German Army Inventory of Serbian Small Arms 1941				
No. D.50/1 01(j)	German Name	Yugoslav Name	Calibre	Origin and Model
491/4	Karabiner 491/4(j)	Karabin M99 C	7.92 x 57 mm	Yugoslav carbine M1899S 7.9 mm repaired Serbian Mauser rifle M1899 7 mm
492	Karabiner 492(j)	Karabin M98	7.92 x 57 mm	German Mauser carbine M98a-M98 AZ 7.9 mm

THE FIRST PARTISAN'S ARMS AND AMMUNITION FACTORY, UZICE
German forces entered the Uzice complex at 2:30 AM on 15 April 1941 beginning Yugoslavia's military industry's period of wartime turmoil. Relying on a report by a Major Rudelsdorf from the *Wehrwirtschafts und Ruestungsamt OKW* (Wehrmacht Military Economy and Armament Management Department for Military Industry of the German High Command) four days later in Vienna, Deputy Ruetter of *Wehrwirtschafts und Ruestungsamt OKW* presented an initial assessment as to how Germany might exploit Yugoslavia's economy, particularly its military industry. On 28 April 1941, the German High Command established the Headquarters for Military Industry, later known as the Military Industry Headquarters for the Southeast, specifically for managing this effort. Within a day, its officials began searching for factories capable of integration into German war industry. They quickly assessed the Fabrika Orugia e Municia, Uzice (the Arms and Ammunition Factory Uzice), and by May decided to shut down the factory for the moment. The factory was immediately closed and guarded.

At the same time, the Germans inspected the thousands of Model 1898 7.92 mm disassembled barrels stored there. They determined that the barrels were of secondary importance and directed they simply be moved to a collection point for seized arms and military equipment, the *Kriegsbeutesammellager* located within the barracks of the 2nd Battalion of the 4th Regiment "Stefan Nemanja." This site was under the direct control of the *Feldzeugstab* 26 (Ordnance Staff 26) at Pozega.

Serbian and Yugoslav Mausers

The decision to close the Uzice factory was based on the fact that all of the manpower available to the Germans was being used to operate the Military Technical Institute at Kragujevac, the Obilcevo and Ravnjak factories, the VTZ factory at Cacak (a private concern known as VISTAD for Valjevo, Visegrad and Baric), and other industries related to aircraft production, such as the Kraljevo factory, Filips-radio, Teleoptic, Nestor A.D, and Miron. Consequently, when the Partisans saw that the Uzice factory was being neglected, they moved quickly to reoccupy it on 24 September 1941.

Partisan commanders could hardly believe that they had come into possession of a practically undamaged and complete manufacturing plant. In a report dated 10 October 1941, the General Headquarters of the National Liberation Partisan Detachments of Yugoslavia (VS NOPOJ) noted the "enormous quantity of arms, ammunition and raw materials necessary for war material production found at Uzice," and stressed the factory's importance to future military operations.

THE *PARTIZANKA* MAUSER

Within the Krcagovo factory complex, Partisans found some 15,000 Gew.98 rifle barrels in 7.92 mm calibre and the same quantity of Model 1898 stocks already adapted to Model 1924 specifications. They also found some 50 Model 1924 barrel-sight assemblies. The Germans may not have had an immediate need for this materiel but the Partisans did as their military forces suffered from a chronic shortage of weapons. They urgently needed the potential rifles represented by the barrels, sight assemblies, stocks, and other parts in the factory.

In the initial stages of organizing to reestablish full production, Misha Savatich, a former engineer originally assigned to an auxiliary plant in Banovina, was brought in to run the main factory complex at Uzice. He was replaced a short time later by a collective body, in effect, a political department de jure in the best communist tradition. But, in all professional and operational matters the factory was run by Nikola Zlatkovich, a former Kragujevac employee.

Converting the recaptured materiel promised to be considerably more difficult than it had been under peacetime production. Standards had to slip and one of the first expedients decided on was to add the sporting-arms production line to the work on military weapons. All non-essential steps in production, such as blueing metal parts, were discontinued and arms were assembled only using existing parts. Such work did not demand sophisticated and powerful machinery but could be done manually using relatively simple hand tools. So, practically all the original metal

Serbian and Yugoslav Mausers

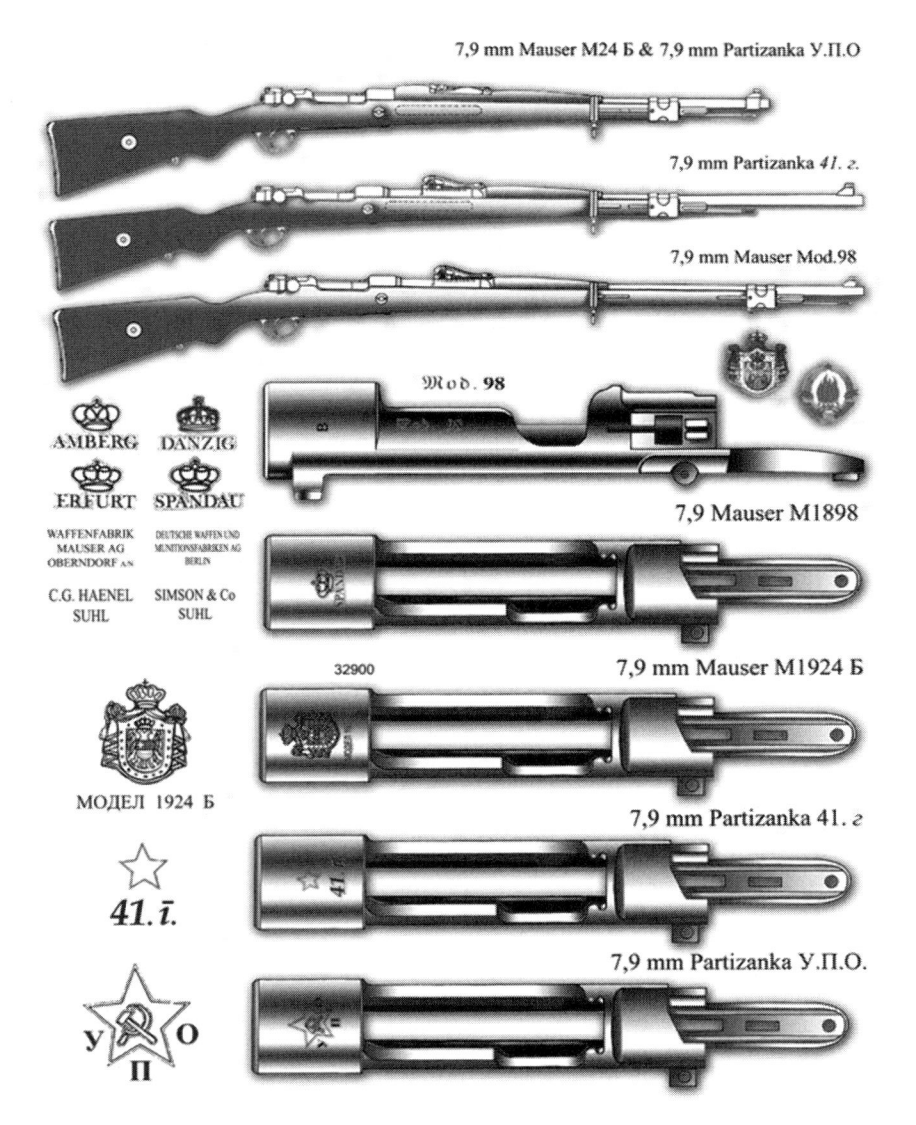

Figure 6-1. Rifles, top to bottom: Yugoslav Model 1924B Mauser 7.92 x 57 mm and "U.P.O." (Uzice detachment) Partisan rifle; *Partizanka* rifle manufactured at FOMU in 1941; German 7.92 x 57 mm Mauser M1898 rifle. Receivers, top to bottom: Original German receiver Model 1898/1917; converted Yugoslav Model 1924B receiver, 1939-1941; *Partizanka* receiver, 1941; "U.P.O." receiver, 1941.

Serbian and Yugoslav Mausers

parts found on Partisan-made Mausers, including the Model 1898 "long" barrels with either "Lange" or "S" sight assemblies, and the receiver, bolt, magazine, trigger mechanism, and bands were simply fitted without any alteration into the converted stocks, see Figure 6-1.

The steel disc remained in the buttstock if the original rifle from which it had been taken was made before 1905 or the domed washer and tube inletted into the buttstock to assist in disassembling the bolt, if made after.

On weapons produced prior to 1905, the original markings remained on the receiver ring and left side including the manufacturer's name, year of manufacture, and the model mark, "Gew.98," as well as the cartridge "S" marking. These weapons also had the rear sights graduated for the "S" cartridge rather than the later Lange Vizier sight.

This mixing of Model 1898 parts and stocks converted to Model 1924 specifications led to the production of some bizarre-looking Mausers, but they had the Mauser's ballistic characteristics, which is what was really important. They were designated *Partizanka* rifles, see Figure 6-2 .

One of their more notable characteristics was how the 740 mm (29.13 inch) long Model 1898 barrel protruded 151 mm (5.94 inches) beyond the Model 1924's stock nose. The Model 1898 upper band with the bayonet lug was also used, although a bayonet could not be mounted for obvious reasons.

"*Partizanka* production" illustrated some of the problems that involved using parts that were forty and more years old. The ballistics of the rifles that resulted fell short of the Model 1924 series of rifles. Part of the problem simply involved the manual fitting of parts, particularly barrels, and a hurried production schedule. Other problems involved the arbitrary selection of the "Lange" or "S" sight assemblies without regard to type of barrel on which they were mounted.

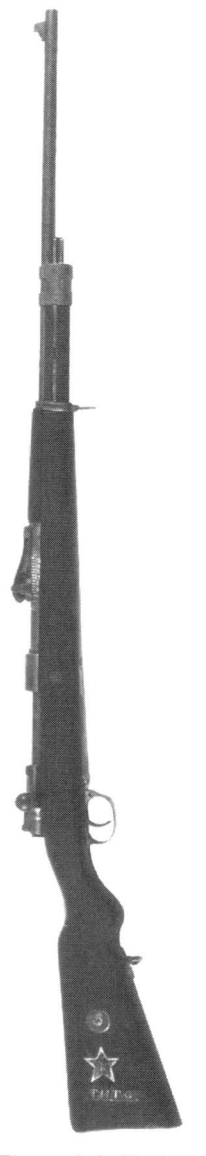

Figure 6-2. Model 1941 *Partizanka* 7.92 x 57 mm rifle.

Serbian and Yugoslav Mausers

Every rifle was test fired at the factory's rifle range under the supervision of "Rifle Controller" Zlatkovich, who was responsible for their inspection and receiving. At the beginning of November 1941, the headquarters staff of the local Partisan detachment and the factory's management decided to send 700 rifles from the first week's production to the Supreme Headquarters for approval. Headquarters personnel expressed their satisfaction with the quality of these weapons but they were dissatisfied with the discipline of the force assigned to protect the shipment. It appears that members of the local unit on their own initiative "acquired" some 252 of the 700 *Partizanka* rifles during the course of the journey in the best military tradition of "scrounging."

The remaining weapons were sent immediately to the Posava Partisan detachment and to the security company attached to Supreme Headquarters. There is some indication that the rifles from this first series were assembled more carefully. The original markings were removed from the receiver ring and replaced by an engraved five-pointed star with hammer and sickle stamped below the year of production, that is: "41. g" for 1941, refer to Figure 6-1. This conclusion is based on one surviving example (serial number 18,365) from the 448 *Partizanka* Mausers delivered to the Partisan security force at Supreme Headquarters. This weapon had "S" sights, Prussian gothic proof marks, and the markings on the receiver ring as described above. This rifle is now in the possession of the Museum of Yugoslav History, formerly "The 25th May Museum," in Belgrade.

Another group of forty rifles showing workmanship equal to that of the prewar Model 1924Bs were assembled in the middle of November at the First Partisan's Arms and Ammunition Factory, Uzice. These rifles were intended for select members of the Supreme Headquarters and the factory's management. These arms were made up using Model 1924B parts and the correct sight assemblies from the last unfinished batch at Kragujevac which were awaiting assembly when the Germans invaded in April 1941.

On the receiver ring of these exemplary pieces both the Kingdom of Yugoslavia coat of arm and designation "M.24.B" were removed and replaced by an engraved five-pointed star with hammer and sickle and the initials of the Uzice Partisan detachment "U.P.O." (in Serbian, "У.П.О."), refer to Figure 6-1. In his diary, Vladimir Dedijer stated that one of these rifles was taken in 1944 by a Yugoslavian military delegation as a gift to Moscow. Another of this group of forty rifles is presently in the Serbian National Museum at Uzice. The rifle (catalogue number 305) has all characteristics of prewar Model 1924B but is recognizable

Serbian and Yugoslav Mausers

by the absence of a handguard. The stock is an altered German stock of the type produced prior to 1905. The serial number of the altered weapon, "V78131," is stamped on the bolt handle. The serial number was probably stamped after conversion as, since 1917, the Germans used serial numbers with the prefix letters from "a to z."

NOTE: Vladimir Dedijer (1914-1990) politician and historian, joined the Partisans as early as 1941.

    This series of *Partizanka* rifles was assembled in the Krcagovo factory just before German air raids began on 28 September 1941. The first attacks caused little damage but a bombing raid on 19 October was disastrous and compelled the factory to relocate to safer sites.
    The rifle assembly lines were quickly moved to three nearby locations, a transfer made easier since the "machinery" consisted mostly of hand tools and accessories. All of the rifle parts apparently remained in storage at the now-deserted factory and were only transferred to one of the three assembly points as required. This assumption is based on the fact that the Germans found a large number of rifle barrels in the extensively damaged Krcagovo factory complex when they reoccupied Uzice on 29 November 1941. Until then, the dispersal of rifle production reduced the effects of the punishing air attacks and other actions, particularly that of an explosion (presumed to be sabotage) that occurred on 22 November.
    Operations continued until the city fell a week later. The number of *Partizanka* rifles assembled between 24-26 September to 28-29 November 1941 differs from one account to another. Intelligence data from the headquarters of the Domobran Ministry of the Independent State of Croatia shows that from November 1941 on, the daily capacity of the factory was 250 rifles. From this, it would have been possible for the Partisans to have remanufactured as many as 15,750 rifles, making the estimate of 10,000 *Partizanka* rifles believed to have been produced reasonably accurate.

CHAPTER 7
POST-WORLD WAR II YUGOSLAVIA

In the years immediately after the Second World War, Yugoslavia's army faced a shortage of even the most basic infantry weapons. It also lacked a credible anti-armor capability. This situation profoundly influenced the domestic military industry to concentrate on developing a strong rifle, submachine gun, and shoulder-fired antitank weapon capability; in addition, it concentrated on converting captured weapons.

THE MAUSER 7.92 MM MODEL 1898N

During 1945, the rifle and machine-gun workshops of the Military Technical Institute at Kragujevac, now renamed VTZ 21st October factory, were reconstituted and quickly managed to repair and make serviceable some 100,000 rifles of different patterns. On 25 June 1946, Yugoslavia's Military Industry Directorate informed the General Staff that Kragujevac had the ability to manufacture, among other weapons, 120,000 rifle barrels and repair 23,000 rifles. Consequently, on 17 July, the General Staff decided to concentrate weapons production in that factory and ordered that in 1947 "an effort should be made" to direct all orders for firearms to the VTZ 21st October facility.

The General Staff's order was, however, in conflict with the military's desire to have the country's main center for firearms production concentrate on 7.92 x 57 mm calibre Mauser weapons. The army's position prevailed and on 8 August 1946 the General Staff directed the VTZ 21st October factory to only overhaul and repair 7.92 mm Mauser rifles. The General Staff went on to calculate, based on the government's Military Industry Directorate's optimistic projection, that Yugoslavia's arms and shipbuilding capability for the 1947 to 1951 period could produce an unrealistic 10,000 7.92 mm Mauser rifles in 1949, 50,000 rifles in 1950, and 120,000 rifles in 1951!

The manufacturing capability for such new weapons simply did not exist, however, and actual production figures are shown in Table 7-1.

There was more behind the report than a frank assessment of domestic capabilities. Yugoslavia's army possessed large quantities of captured German army Mausers. Using Yugoslavian army nomenclature, the quantity of each type of rifle and its country of origin are shown in Table 7-2 and Figure 7-1. Figures 7-2 and 7-3 show the most common of the various markings on the captured German rifles before and after

Serbian and Yugoslav Mausers

refurbishment. Many, but not all, of the captured German rifles can also be identified by their straight bolt handles, see Figure 7-4.

Table 7-1 Rifle Production* from 1947 to 1952				
Year	Model			
	M24/47 7.92 x 57 mm	7.92 x 57 mm M98n** (M98/48)	7.92 x 57 mm M24/52	7.92 x 57 mm M48
1947	10,935			
1948	?	53,776		
1949		50,000		
1950		30,165		53,790 (52,002)*
1951		187,599		91,086 (92,037)*
1952			197,599	94,476 (94,874)*

* Totals of rifles manufactured according to the archives of the Museum, Old Foundry, Zastava Arms Factory, Kragujevac. The numbers in brackets are from data in the Military Archives, Belgrade.
** "N" indicates that the rifle was originally manufactured as a German Army rifle.

Table 7-2 1013 – Rifles and Carbines, Inventory*			
Ordinal No.	Name	Origin and Model	Storage Number
114	Puska 7,9 mm M24 The M24 rifle 7.9 mm	Yugoslav M24 Mauser 7.92 x 57 mm	1013-1023-144
115	Puska 7,9 mm M24 (b) Rifle M24(b) 7.9 mm	Belgian M24 Mauser 7.92 x 57 mm	1013-1023-145
116	Puska 7,9 mm M24 (c) Rifle M24(C) 7.9 mm	Czech Mauser rifle vz.24 7.92 x 57 mm German Gewehr 24(t) & Rumanian variant ZB Md.24 7.92 x 57 mm	1013-1023-146

Serbian and Yugoslav Mausers

Table 7-2, cont. 1013 – Rifles and Carbines, Inventory*			
Ordinal No.	Name	Origin and Model	Storage Number
123	Puska 7,9 mm M98 (n) Rifle M98(n) 7.9 mm	German Mauser carbine K98k 7.92 x 57 mm (German marking: Mod.98)	1013-1023-147
124	Puska 7.9 mm M98 (n), sa dugom cevi Rifle M98(n) 7.9 mm with long barrel	German Mauser rifle M1898 7.92 x 57 mm (German marking: Mod.98)	1013-1023-0148

* Data taken from *Imenici pesadijskih sredstava* (Infantry Arms and Armament List) Ministry of Defense, Belgrade, 1966 and 1984. This document was classified top secret at the time.

The Serbian army's traditional allegiance to Mauser firearms was inherited by its successor, the Yugoslavian army, along with a preference

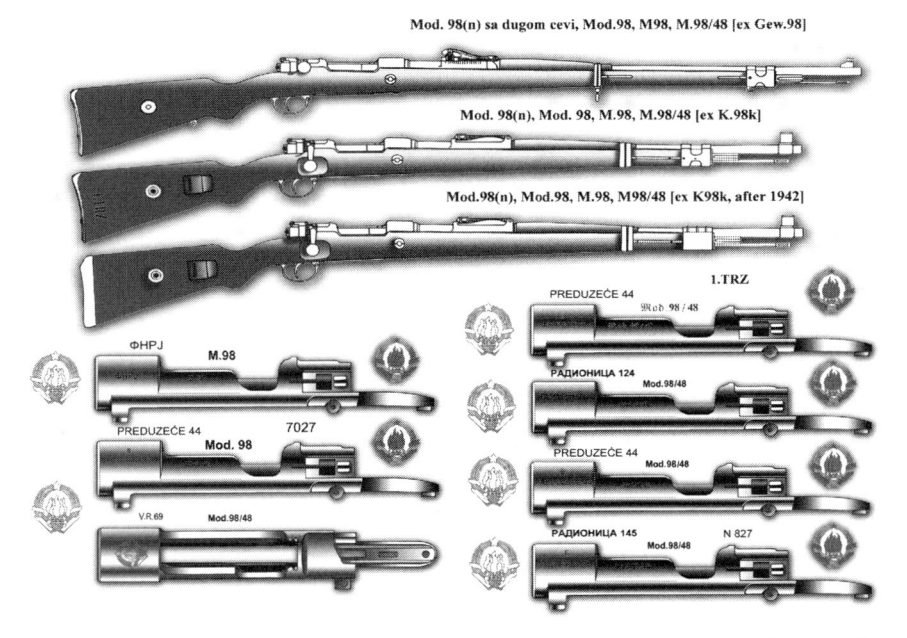

Figure 7-1. Various German Mauser rifles which were converted for use by Yugoslavia after World War II.

Serbian and Yugoslav Mausers

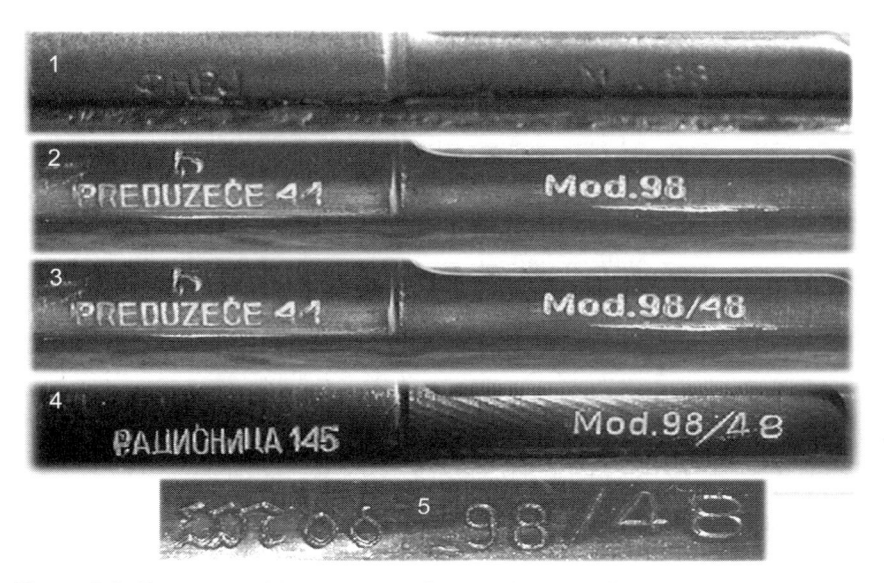

Figure 7-2. Factory markings on the left side of altered German M98 and M98/48 receivers (top to bottom): 1) FNRJ (Federal National Republic of Yugoslavia); 2 & 3) PREDUZECE 44 (Kragujevac); 4) RADIONICA 145 (location unknown); 5) original German model marking with the added Yugoslav "48" suffix (after 1950).

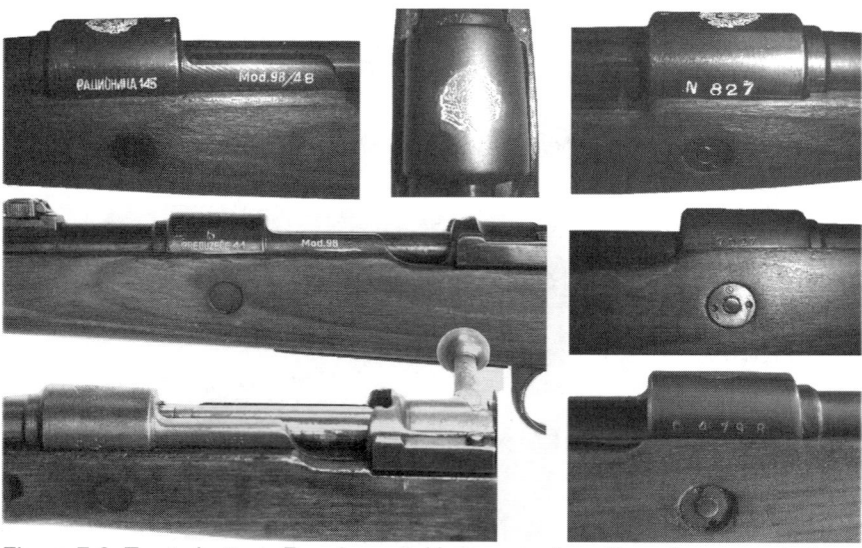

Figure 7-3. Top to bottom: Receiver markings on various Yugoslav-converted German Mauser rifles.

Serbian and Yugoslav Mausers

Figure 7-4. Close-up of a German Model 1898 rifle converted to the Yugoslav Model 1948 pattern.

for "universal" ammunition that could be used for a variety of weapons. Army leaders were particularly impressed with the very effective MG34 and MG42 German machine guns that also used the same cartridge as the rifle carried by Wehrmacht soldiers. Officials of the Military Industry Directorate simply assumed from the domestic industry's prewar success with the Model 1924 rifle as produced by VTZ Kragujevac, and with 7.92 x 57 ammunition production, also at VTZ Kragujevac as well as FOMU, Uzice, and VISTAD, Valjevo, that its arms industry was immediately capable of full production. They ignored the fact that after four years of bitter fighting, Yugoslavia lacked the experienced staff, technology, and raw materials that had made its arms program so successful before the German invasion. Government officials also ignored or were not aware that some of the Model 1924's technical documents necessary for production had also been destroyed in the fighting.

THE MAUSER 7.92 x 57 MM MODEL 1924/47
Repair and refurbishment of the captured Wehrmacht 7.92 mm Mauser rifles began in the second half of 1946. Throughout 1947 the Red Flag Enterprise factory—VTZ Kragujevac was renamed the Red Flag Enterprise (Preduzece Crvena Zastava) on 13 January 1948—working to a quota of 23,239 finished weapons, assembled 10,935 new 7.92 mm Model 1924/47s (as the converted weapons were designated) from the metal parts of the old weapons and from either newly manufactured or

Serbian and Yugoslav Mausers

repaired stocks. These stocks had the identical characteristics of prewar Model 1924 rifle stocks. A year later, factory workers under a quota of 114,271 weapons converted 53,776 Model 1898 rifles and K98k carbines into M98(n). The suffix "n" denotes a "German rifle"—*nemacka puska*. The original German model markings were retained on the left side of the receiver or else replaced with a new Yugoslavian marking reading "M.98," refer to Figure 7-1.

Concurrent with this effort, the technical maintenance works and the military technical workshops were busy assembling Model 1924/47 Mausers and converting Model 1898 Mausers into the Model 1898(n) version. It is not known how many weapons these two shops finished. The extensive efforts by factory employees made it possible for the army's General Staff to provide a better distribution of armament on 19 March 1948 than originally believed possible, see Figures 7-5 and 7-6.

Figure 7-5. A preconscription training enrollee with the M24 rifle in 1953.

The repaired rifles and carbines were issued to troops stationed in districts I, II, III, V, and VII of Yugoslavia. There were some exceptions to this distribution. Soldiers of rifle regiments, reconnaissance companies, and independent battalions in districts II and VII, soldiers in the independent artillery brigades in districts III and V, soldiers assigned to independent engineer regiments and battalions, the independent communication regiment and other attached units in districts III and V, and soldiers assigned to the KNOJ (Corps of National Defense of Yugoslavia), the Yugoslavian navy, tank units, and mechanized units did not received the weapons. The remaining 7.92 mm Model 1924/47 rifles were retained in case the reserves were ever mobilized.

Serbian and Yugoslav Mausers

Figure 7-6. Young enrollees in preconscription training in 1953. They are armed with Model 1924 and Model 24/47 rifles.

Serbian and Yugoslav Mausers

During 1949, the "Red Flag Enterprise" factory converted around 50,000 Mausers to the Model 1924/47 configuration, see Figures 7-7 and 7-8. Figure 7-9 shows the various die stamps used to mark the model number on the left side of the receiver. Figure 7-10 illustrates the receiver ring marked with the newly adopted, postwar national crest and the serial numbering on the right side of the receiver ring. See also Table 7-3 for a complete review of Model 1924 rifle markings.

The Model 1924/47 7.92 x 57 mm rifles were made from Model 1924 rifle parts and carbine stocks. For the sake of uniformity, on stocks initially intended for carbines, the holes for the rear sling swivel mounted on the side of the stock were filled in.

	See Also Table C-1, C-2, C-3, C-4, C-5, C-6, C-7, C-8	Coat of Arms	Pattern Mark	Manufacturer's Markings	Country's Name	Year of Production	Ruler's Monogram	Full Serial Number	Last Three Digits of Serial Number	Smokeless Powder Proof Mark	Inspection Mark	Acceptance Mark
Table 7-3 Mauser Rifle M24/47 7.92 mm												
Barrel:								x / y		x / y	x / y	x / y
Top	x / y	x										
LEFT			y	x						x / y	x / y	
Receiver												
Right								x / y				
Left				y								
Recoil Lug											x / y	
Bolt Assembly												
Bolt Body												

Serbian and Yugoslav Mausers

Table 7-3, cont. Mauser Rifle M24/47 7.92 mm										
See Also Table C-1, C-2, C-3, C-4, C-5, C-6, C-7, C-8	Coat of Arms	Pattern Mark	Manufacturer's Markings	Country's Name	Year of Production	Ruler's Monogram	Full Serial Number	Last Three Digits of Serial Number	Smokeless Powder Proof Mark	Inspection Mark	Acceptance Mark
Bolt Handle Base										x y	
Bolt Handle Neck							x y				
Bolt Handle Ball									x y		
Cocking Piece										x y	
Bolt Sleeve with Gas Shield								x y			
Safety Lever								x y			
Extractor											
Extractor Collar											
Ejector											
Bolt Stop										x y	
Trigger											

Serbian and Yugoslav Mausers

Table 7-3, cont. Mauser Rifle M24/47 7.92 mm										
See Also Table C-1, C-2, C-3, C-4, C-5, C-6, C-7, C-8	Coat of Arms	Pattern Mark	Manufacturer's Markings	Country's Name	Year of Production	Ruler's Monogram	Full Serial Number	Last Three Digits of Serial Number	Smokeless Powder Proof Mark	Inspection Mark	Acceptance Mark
Sear-fork										x y	
Trigger Assembly											
Trigger Guard Plate							x y				
Magazine Floor Plate							x y				
Stock		y						x y			
Butt Plate								x y			
Front Band, Lower Band										x y	
X = Type 1 (for military use) Y = Type 2 (for export)											

THE MAUSER 7.92 MM MODEL 1924/52-Č

During 1952 when the technology and production lines were being developed to produce the Model 1948A rifle production, the factory of Kragujevac begun repairing the Czechoslovakian Mausers vz.24s in 7.92 x 57 mm calibre.

Germany had occupied Czechoslovakia in March 1939 and added many existing Czech small arms to their armories. Germany therefore included some ninety percent of seized Czech vz.23 & vz.24 rifles as part of the reparations owed to Yugoslavia, see Figure 7-11. These arms retained their original markings as follows:

Serbian and Yugoslav Mausers

Model vz.23:

ČS ZÁVODY NA/VÝROBU/BRNO

with pattern mark "vz.23" and four-digit serial number with preceding letter (K5555).

Model vz.24:

ČESKOSLOVENSKÁ ZBROJOVKA/ BRNO

or

ZBROJOVKA BRNO, A.S. VZ.24 (ČESKOSLOVENSKÁ ZBROJOVKA, A.S, BRNO)

together with pattern mark "vz.24" and serial number consisting of numerals and letters (5555A3), see also Figure 7-12.

Rifle production which had been contracted by the Czech Army was discontinued under the Nazi regime but production of spare parts was

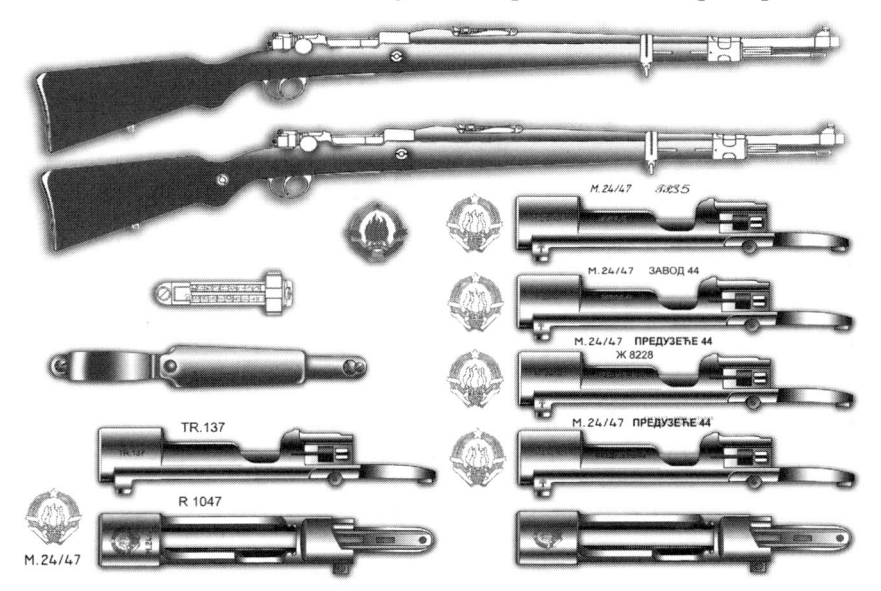

Figure 7-7. Above, Model 1924/47 with the infantry rifle stock; below, the Model 1924/47 with the carbine stock. Various receiver markings are also shown.

Serbian and Yugoslav Mausers

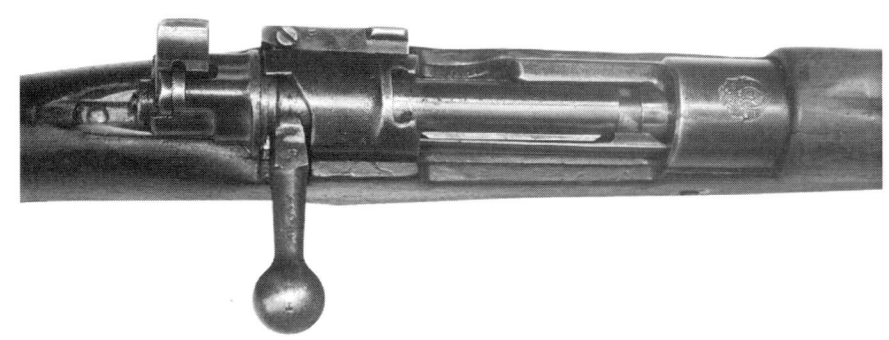

Figure 7-8. The Model 1924/47 receiver and bolt.

Figure 7-9. Factory markings for the Model 1924/47. Top: PREDUZECE 44, Kragujevac, 1948-1960; below, left: ZAVOD 44, Kragujevac, 1945-1948; below, right: TRZ-5 – date unknown.

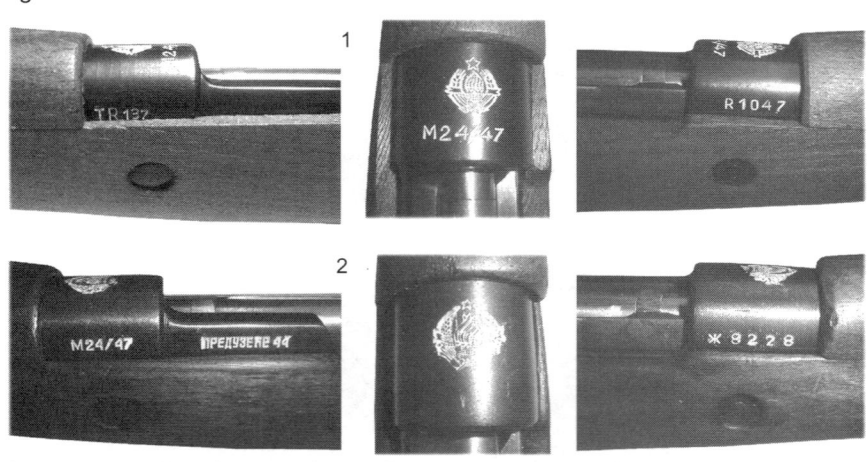

Figure 7-10. Receiver markings for the Model 1924/47: 1) TRZ-137 (date unknown), serial number with Latin letter prefix on the rifles intended for export; 2) PREDUZECE 44 (Kragujevac), serial number with Cyrillic prefix letter on the rifles intended for use by the Yugoslav military.

Serbian and Yugoslav Mausers

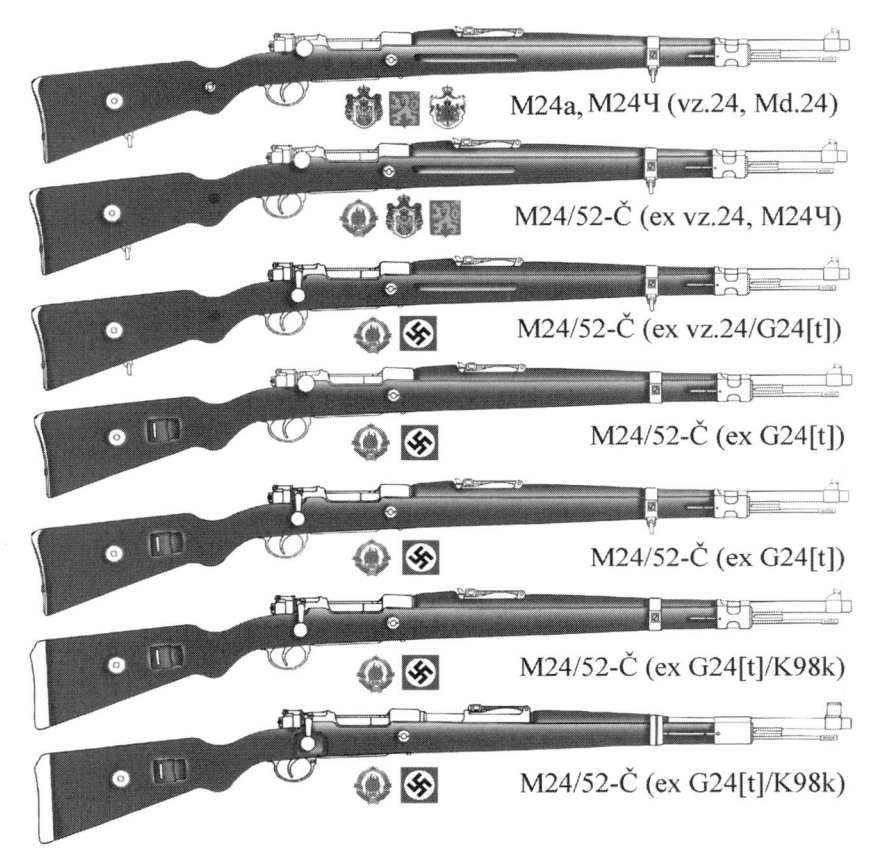

Figure 7-11. Shown here are all variations of the Yugoslav Model 1924/1952 and the Czechoslovakian vz.24 and German G24t and K98k rifles from which they were converted.

continued. The Czech factories were also permitted to fulfill a Rumanian contract for delivery of the Model 1924 7.92 x 57 mm rifles (Rumanian Puşcă ZB Md. 1924). Romania had ordered 700,000 Mauser vz.24 rifles shortly before the Germans marched in Czechoslovakia, of which 445,640 were delivered. The Rumanian factory, Uzunele Metalurgice Copşa-Mică-Cugir, never did manage to master the production of the Model 1924 rifle.

During 1941, the Waffenwerk Brünn AG–Werk Považka, Bystrica factory was ordered by the German occupation forces to continue the stan-

Serbian and Yugoslav Mausers

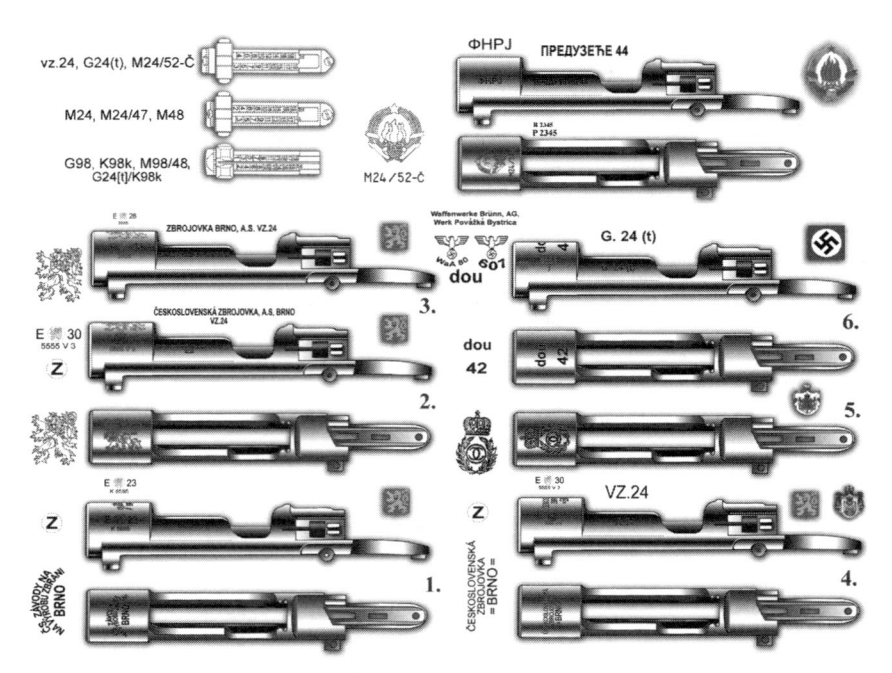

Figure 7-12. Receiver markings and rear sights for the Yugoslav Model 1924/52-Č rifle: 1-6, before conversion showing original markings; 7-8, after conversion showing Yugoslav markings.

dard 7.92 x 57 mm vz.24 Mauser rifle production using available spare parts. These rifles were designated by the Germans as the G.24(t).

As an aside, this factory was actually founded in the city of Brno as early as 1919 as the Ceskoslovenske Zavody na Vyrobu Zbrani factory. Two years later a new factory section was put into operation and named Ceskoslovenska Zbrojovka Povázká Bystrica which, together with the main plant in Brno, formed the company Ceskoslovenska Zbrojovka akc. Spol. Brno.

The first batch of G.24(t) rifles differed in minor details from the pre-war vz.24. The original solid walnut stocks with forged, flat butt plates were used instead of the later laminated beech stock and stamped cupped butt plates. The stock had a finger groove in either side of the forend. A straight bolt handle was employed as was the earlier rear sight graduated 300 to 2000 meters (328 to 2,187 yards).

Serbian and Yugoslav Mausers

Only the bottom rear sling swivel was used. The rear sling swivel normally mounted on the left side of the buttstock was eliminated and its slot was filled with a wood insert.

Later-production G.24(t) rifles were made with the laminated beech stock and cupped butt plate. The rear sling swivel on the bottom of the stock was eliminated and a slot cut through the buttstock to attach the sling in the same manner as on the German K98k rifle.

At the same time the original Czech rear sights graduated from 300 to 2000 meters (328 to 2,187 yards) were retained. On the upper side of the receiver ring a new code mark "dou" and the year of manufacture were stamped, while the model marking "G24(t)" was stamped on the left side. Most parts were stamped with "WaA Považke Bystrice" followed by the numbers "80," or "607," although the old Czech stamp "BZ" inside a circle can also be found.

As supplies of Czech straight bolt handles were lacking, the turned-down Model 1898 bolt handles were used to complete a number of the G.24(t) rifles. In order to fit the bolt handle, a small cut compatible with neck of the bolt handle diameter was milled into the stock. Moreover, at the beginning of 1942, as a result of an increasing shortage of arms the Germans ordered the factory to start assembling the rifles that combined existing Czech and new K98k parts. These included 1) German-made laminated wood K98k stocks, 2) German barrels manufactured at Steyr bearing the stamps "234 4 c BO" and "WaA 77," 3) German butt plates, 4) bands, 5) bayonet lugs stamped "gqm" ("Loch und Hertennberger, Idar-Oberstein") and "WaA 77" and "623" ("Steyr-Daimler-Puch AG," Werk Steyr) and 6) German turned-down bolt handles.

The receivers, magazines, and magazine floor plates (bearing the stamps "WaA 80") with their original markings in Czech were taken from the Czech plant. Finally, the rear sights of the K98k were added. By the end of 1942 both Czech plants had stopped producing G.24(t) rifles and assembling combined weapon which the Germans designated vz.24/K98k. They were now able to devote their production resources to manufacturing the German K98k which can be identified now with the markings "dot" or "swp" (Brno Werke) or "dou" (Povazka Bystrica). Refer to Figure 7-12 for markings that identify the various models.

During 1944, the Yugoslav army seized a large quantity of vz.24, vz.24/K98k, G.24(t), as well as ZB Md.1924 rifles. More were acquired after the war ended in May 1945 as war reparations as well as a large quantity of prewar Yugoslav 7.92 x 57 mm Model 1924A (Model 1924C, vz.24) rifles. The stocks on these rifles had the slot for the rear side sling swivel filled in. They were also equipped with the barrel bands typical

Serbian and Yugoslav Mausers

Figure 7-13. Model 1924/52-Č markings in Yugoslav military service. The original serial number with a Latin letter prefix was replaced by a Cyrillic letter "P" (Latin R).

of the Czech carbines. They can be identified by the markings on the receiver ring which included the coat of arms of the FNRJ and the model marking "M.24/52-Č." The suffix Č=Ch denoted Czech manufacture. Depending on its condition when it reached Yugoslavia, it either retained its original barrel or had it replaced with a new domestically manufactured Model 1948 barrel. See Figure 7-13 for markings.

Since the repaired rifles came from different sources, other variants of Model 1924/52-Č than those identified to date and listed here may be encountered. The first group of rifles were assembled from original parts of prewar Czech and/or Yugoslav vz.24/Model 1924A rifles; the second group was assembled between 1941-1945. These last incorporated different types of the stocks and straight or turned-down bolt handles. It is impossible to fix the number of repaired arms exactly but, no doubt, it exceeds 50,000. Examples of these later rifles can be seen in Figure 7-14.

Serbian and Yugoslav Mausers

Figure 7-14. Yugoslav Army (YNA) privates equipped with the Model 1924/52-Č rifle.

CHAPTER 8
THE YUGOSLAV MODEL 1948 7.92 X 57 MM RIFLE AND ITS VARIATIONS

In the early 1950s, gunsmiths assigned to sections I and V at the Red Flag Enterprise factory developed a prototype of a new rifle which was designated the Model 1948 7.92 x 57 mm Rifle (M48), see Figure 8-1. Specialists in design, production, and administration such as Jordan Kosanich, Zivadin-Zira Jovanovich, Djura Cvijovich, Djurdje Matich, Borivoje Milenkovich, Ratko Stojanovich, Obren Paunovich, Zoran Mishich, Radivoje Tomovic and Borislav Maksimovich (head of a gunsmith shop section from 1947 to 1949), were assigned to "Institute No. 11" to accomplish this work. In principle, the Model 1948 was based on the prewar Mauser Model 1924 but with some details taken from the German carbine K98k.

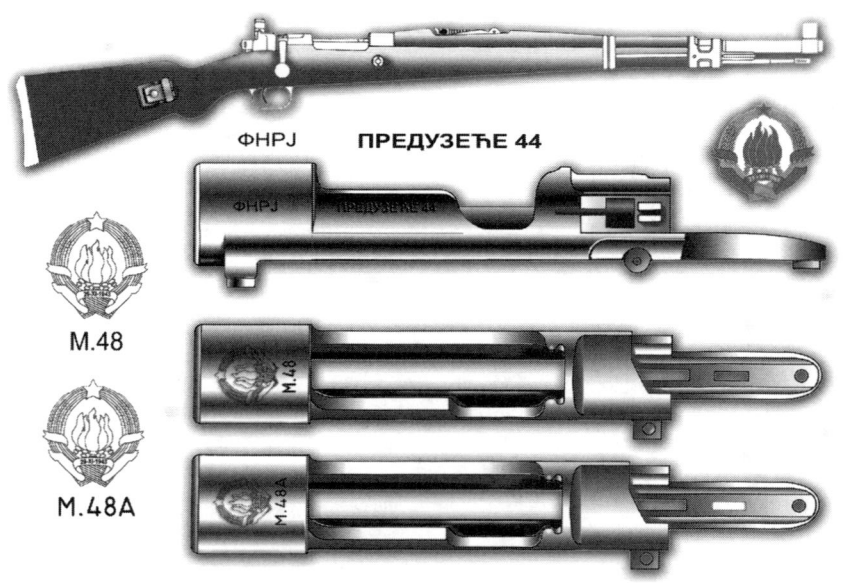

Figure 8-1. The 7.92 x 57 mm Model 1948 rifle and receiver and Model 1948A receiver.

THE MAUSER 7.92 MM M48—DEVELOPMENT AND PRODUCTION

The German K98k receiver bolt head diameter was 35.8 mm (1.41 inches) and its receiver length was 222.25 mm (8.75 inches) compared to the

Serbian and Yugoslav Mausers

Yugoslavian Model 1948 rifle which also had a receiver bolt head diameter of 35.8 mm (1.41 inches) but a receiver length of only 215.9 mm (8.50 inches), the same as the Model 1924 and Model 1924/47 rifles.

The Model 1948 rifle's turned-down bolt handle and sling swivels were based on those of the K98k, but the its bolt length was 6.0 mm (0.236 inch) shorter than the German K98k and its overall travel was 155.32 mm (6.115 inches) compared to the 161.3 mm (6.350 inches) of the K98k.

During the rifle's development, the most serious difficulties encountered had to do with some of the materials used in its construction. The rifle required thermic-treated steel parts but the mechanical characteristics of steel before and after thermic treatment were unknown. Specifically, there was no information available regarding specifications for time, temperature, and the depth of treatment. The answers were only found through extensive trial and error. The most serious problem occurred in the large number of rejects during bayonet production. Not until 1952 did an engineer named Slobodan Mitrovich solve the problem and production reached full capacity.

As a testament to their hard work, factory employees produced 52,002 M48 7.92 mm rifles in 1950; 92,037 in 1951; and 94,476 in 1952. Another account listed the number of weapons produced as 53,790, 91,986, and 94,874, respectively. Nevertheless, by 1953, the factory had delivered to the Army a total of 238,515 new Model 1948 rifles and 343,710 repaired 7.92 mm Mauser M98 and K98k (bearing the new model markings, "M98/48"), including the Model 1924/47.

The Mauser 7.92 mm M48A, M48B and M53

In 1952, engineers from this factory simplified rifle production when they introduced the technology of stamped parts which was first used in the manufacture of magazine floor plates. The move to this technology was driven by a new generation of engineers such as Djukich, Radmilo Vaskovich, Jovan Ignjatovich, Branko Nikolich, Ratko Stojanovich, Vladan Dimitrijevich, Dobrivoje Jovanovich as well as Jevrosim Popovich, Ljubisha Vuchkovich, Dragi Mihailovich, and Predrag Mirchich.

After installation of the new production line, Kragujevac began serial production of the Model 1948A: 93,091 rifles in 1953, 93,091 (or 103,232 by another count) during 1954, 103,474 (or 104,535) in 1955, and 40,036 rifles in 1956. The drop in production in 1956 stemmed from the switch made to the manufacture of the Model 1948B variant which

Serbian and Yugoslav Mausers

replaced a number of milled parts with parts stamped from steel, including the butt plate and trigger guard, see Figures 8-2, 8-3 and 8-4.

Note: The M48B rifle retained designation M48A on the receiver ring.

So confident was the Military Council of the improved productive capacity of the Kragujevac factory that on 12 December 1956, they initiated a gradual withdrawal of the refurbished Models of 1924/47, 1924/52-Č, 1898, and 1898/48 rifles and their replacement by the domestically manufactured Models of 1948, 1948A and Model 1948B.

New ammunition pouches and cleaning kits were also designed and issued for the Model 1948 series rifles, see Figure 8-5 and refer to Figure 4-18. Made of leather with a smooth finish, they included a single and double pocket pouch in both the Type I and Type II series, and a three pocket pouch in the Type II series. All were marked with the Cyrillic inspection stamp, "BK" (in the Latin alphabet, "VK" for *Vojna Kontrola* or Military Control) in a circle. This stamp was in use from 1947 to 1970. It was worn on the soldier's leather belt by means of two loops attached to the rear panel. The flaps covered the mouth of the pouch to keep out dust, rain and snow and were fastened closed with a leather strap that slipped over a stud sewn and riveted to the bottom surface of the pouch. The pockets were sized to hold two clips of 7.92 x 57 mm ammunition.

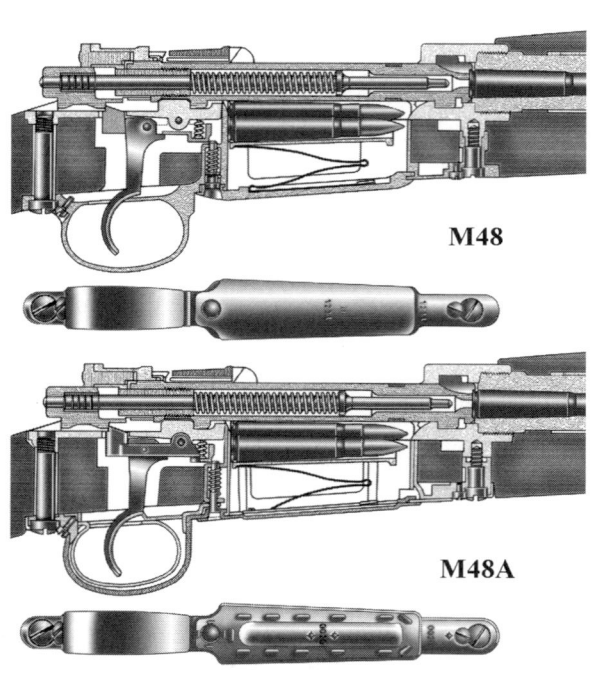

Figure 8-2. Cross section views of the Model 1948 and Model 1948A receivers. Also shown, the floor plates for both models.

Serbian and Yugoslav Mausers

Figure 8-3. The Model 1948A 7.92 x 57 mm Mauser Rifle.

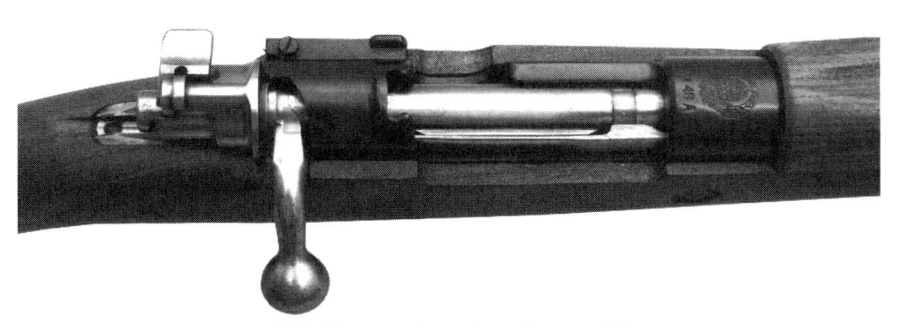

Figure 8-4. The Model 1948A bolt and receiver ring marking.

Two types of grenade launcher were developed for the Model 1948 rifle series, the Model 1951 Energa and the Model 1960 Zastava. Both were of the straight-tube variety and somewhat similar to the American M7 series or the French Strim. They slid partly over the muzzle and were fastened in place with hinged steel straps that were secured with a wing nut, see Figure 8-6.

A special blank cartridge was used to fire the grenades, which were inserted into the launcher. Aiming and ranging was accomplished by a hinged sight that could be raised or lowered. The Model 1951 sight was graduated to 100 meters (109.4 yards) but the Zastava sight was graduated to 300 meters (328.1 yards).

Between 1958 and 1968, Kragujevac's Zastava factory produced 737,486 7.92 mm M48A & M48B rifles, see Table 8-1. Consequently, by 1 April 1967, the Yugoslavian army possessed 932,108 7.92 mm rifles, see Figure 8-7. These included the Models of 1948, 1948A, and 1948B rifles, all in service. The Models of 1924/47, 1924/52-Č, 1898 and 1898/48 were held in reserve, see Table 8-2.

Serbian and Yugoslav Mausers

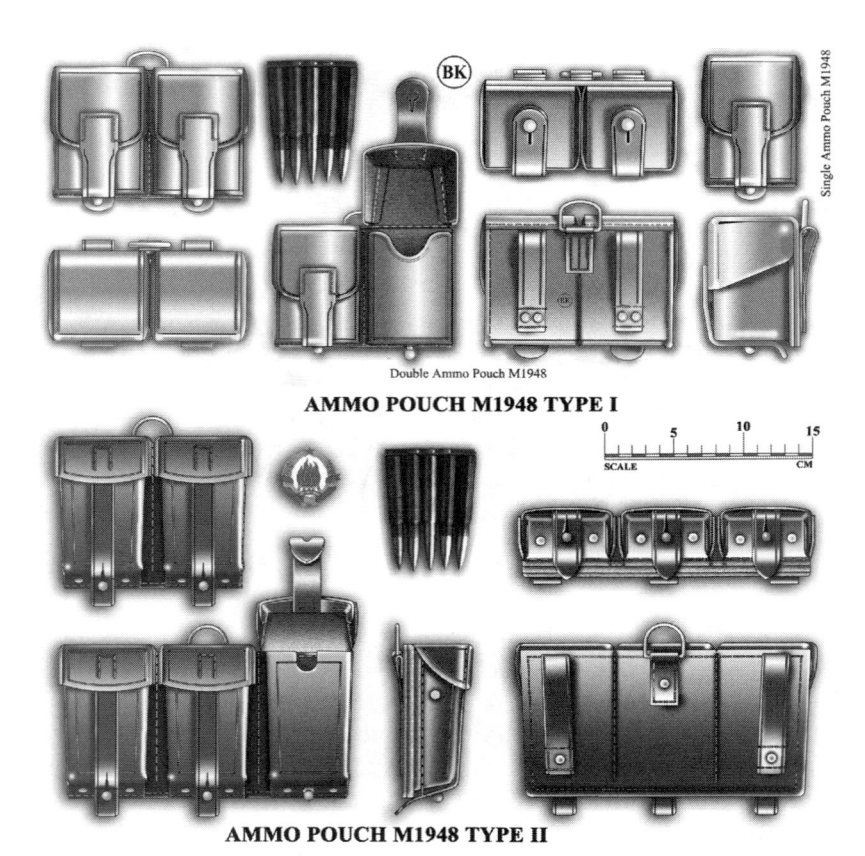

AMMO POUCH M1948 TYPE I

Double Ammo Pouch M1948

Single Ammo Pouch M1948

AMMO POUCH M1948 TYPE II

Figure 8-5. Ammunition pouches issued for the Model 1948 7.92 x 57 mm Mauser rifles.

Table 8-1 Rifle Production from 1951 to 1967*										
Year	Model									
	7.9 mm M24/47	7.9 mm M98n (M98/48)	7.9 mm M24/52	7.9 mm M48	7.9 mm M48/52	7.9 mm M48A	7.9 mm M48B	7.9 mm M48BO	7.9 mm M53	Grenade Launcher M60
1951	?	197,599		92,037 (91,986)						

Serbian and Yugoslav Mausers

Year	7.9 mm M24/47	7.9 mm M98n (M98/48)	7.9 mm M24/52	7.9 mm M48	7.9 mm M48/52	7.9 mm M48A	7.9 mm M48B	7.9 mm M48BO	7.9 mm M53	Grenade Launcher M60
1952			94,476 (94,874)							
1953					4,618	93,091				
1954						93,091 (103,232)				
1955						103,474 (104,535)				
1956						40,036				
1957							351,464	386,022		
1958										
1959										
1960										
1961									100	
1962										
1963										
1964										
1965										
1966									20,100	
1967										

Table 8-1, cont. Rifle Production from 1951 to 1967*

* The number of manufactured rifles is given according to the archives of the Museum, Old Foundry, Zastava Arms Factory, Kragujevac. The numbers in brackets are from data in the Military Archives, Belgrade. Serial numbers did not have prefixes but may show a Cyrillic batch letter.

Ordinal No.	Name	Origin and Model	Storage Number
114	Puska 7,9 mm M24 The M24 rifle 7.92 mm	Yugoslav M24 Mauser 7.92 x 57 mm	1013-1023-0144

Table 8-2
1013 – Rifles and Carbines, April 1967 Inventory

Serbian and Yugoslav Mausers

	Table 8-2, cont. 1013 – Rifles and Carbines, April 1967 Inventory		
Ordinal No.	Name	Origin and Model	Storage Number
115	Puska 7,9 mm M24 (b) Rifle M24(b) 7.92 mm	Belgian M24 Mauser 7.92 x 57 mm	1013-1023-0145
116	Puska 7,9 mm M24 (c) Rifle M24(c) 7.9 mm	Czech vz.24 Mauser 7.92 x 57 mm, German "Gewehr" 24(t) & Rumanian variant "ZB" Md.24 7.92 x 57 mm	1013-1023-0146
123	Puska 7,9 mm M98 (n) Rifle M98(n) 7,92 mm	German Mauser Carbine K98k 7.92 x 57 mm (original German marking – Mod.98)	1013-1023-0147
124	Puska 7,9 mm M98 (n), sa dugom cevi Rifle M98(n) 7.9 mm with long barrel	German Mauser Rifle M1898 7.92 x 57 mm (original German marking – Mod.98)	1013-1023-0148

In the 1960s, the Infantry Department requested development of a lighter version of the Model 1948. Lieutenant Colonel Todor Cvetich designed the variant on behalf of the Institute of Armament and the government gave the project to the Kragejuvac factory, now referred to as Preduzece Crvena Zastava (Red Flag Enterprise).

The new light rifle, temporarily designated the Model 1953, weighed 20 to 35 percent less (3 kilograms or 6.613 lbs) than the Model 1948A, see Figures 8-8 and 8-9. The rifle had an effective range of 500 meters (546 yards). The receiver (see Figure 8-10) was modified with a new and shorter type of safety catch (arrow A), a lightening cut made in the left side of the receiver (arrow B), and larger extractor (arrow C). Engineers also replaced the classic barleycorn front sight with one with integral protecting ears on either side, similar to that used on the Lee-Enfield series of rifles, see Figure 8-11.

In order to make the new rifle suitable for accurate shooting, it was chambered for standard ammunition 7.92 x 57 mm, which, in this lightweight rifle, recoiled strongly. To help tame it, a recoil compensator was attached to the muzzle, see Figure 8-12.

Serbian and Yugoslav Mausers

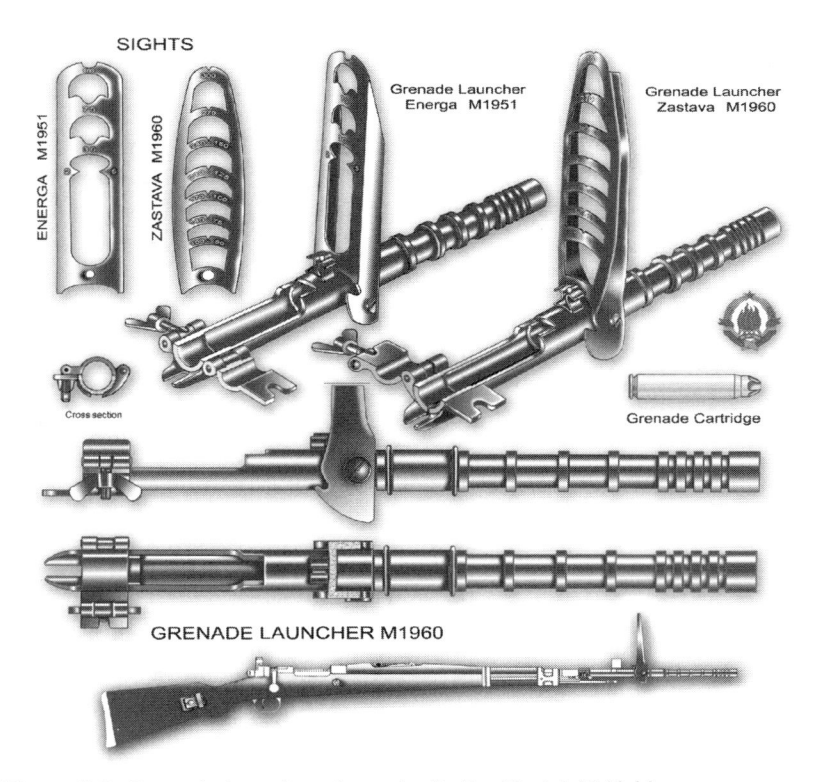

Figure 8-6. Grenade launchers issued with the Model 1948 Mauser.

Zastava produced a prototype and a zero ("0") batch of 100 production "light rifles." The rifles were sent for field-testing but the project was canceled when the Army decided to adopt the new 7.62 x 39 mm calibre, *Poluautomaska puska (PAP)* Model 1959, a semiautomatic rifle. The Model 1959 was the Yugoslavian-manufactured version of the Soviet 7.62 *Samozaryadnyi Karabin Sistemi Simonova Obrazets 1945g* or SKS carbine, see Figure 8-13.

A Word about Stock Woods

All Serbian Mausers manufactured in Germany and Austria had walnut stocks. The Yugoslavian Model 1924 used walnut for its stocks as did the altered Turkish rifles. But after World War II, the shortage and expense of walnut forced the use of other woods for stock making. The Model 1924/47 rifles assembled immediately after the war were mainly stocked

Serbian and Yugoslav Mausers

Figure 8-7. Cadets from the NCO school during training in 1973 firing the Model 1948A rifle. Photo courtesy of Dr. Simon Dragovic.

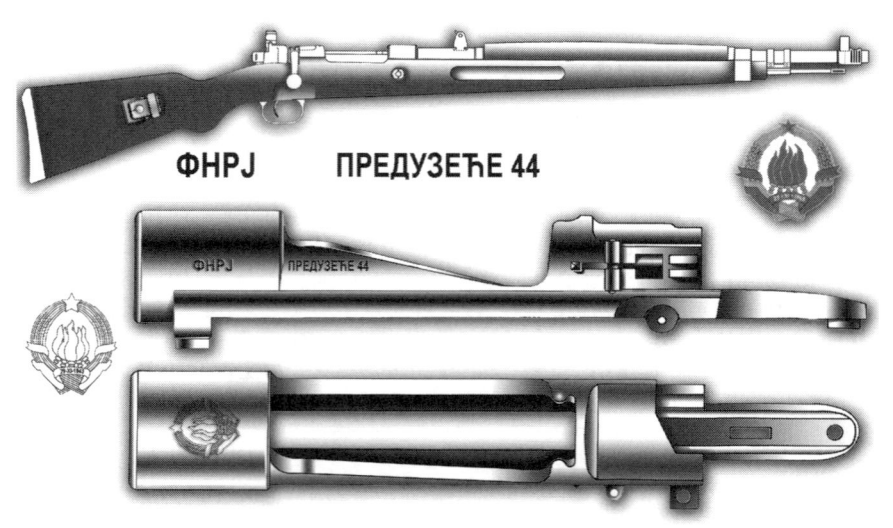

ФНРЈ   ПРЕДУЗЕЋЕ 44

Figure 8-8. The new, lightweight Model 1953 7.92 x 57 mm Mauser rifle.

with old Yugoslav walnut stocks or with newly made walnut stocks. Only rarely were some Model 1924/47 rifles fitted with oak or beech stocks. Repaired German Model Gew.98 and K98k rifles retained their original

Serbian and Yugoslav Mausers

stocks, some of which were made from laminated wood, usually beech.

Postwar rifle developments led to the use of different woods. The Model 1948, Model 1948A and Model 1948B rifles were fitted primarily with beech stocks, although some were also fitted with stocks made from elm. But the entire first series of the Model 1948 rifle had walnut stocks.

The existence of stocks made of teak, or some other atypical wood, confuses the situation somewhat. Yugoslavia exported both completed Model 1948 rifles as well as kits containing only the metal parts. These latter were assembled and subsequently stocked by the purchaser with whatever wood they had on hand or preferred. There is no truth to the rumor that the Chinese firm "Norinco" stocked Model 1948 rifles with teak wood cut in 1996. Other confusing examples result from rifles bought by arms dealers in Canada and the United States of America. While it is certain that Yugoslavia sold completed Model 1924/47, Model 1924/52-Č, Model 1948, Model 1948A, Model 1948B and Model 1948BO rifles with original walnut, beech or elm stocks (the Model 1898/48 rifles had stocks of German origin made of various materials from wartime production, including laminated beech), it is presumed that some dealers replaced damaged stocks with new ones made of different materials or with other Mauser stocks of unknown origin. This is almost certainly the case where dealers bought only the metal components of Yugoslavian-made firearms.

NOTE: "BO" is the abbreviation for *bez oznake* which means "unmarked."

Table 8-3 summarizes the technical specifications for the post-World War II Yugoslav rifle models. Tables 8-4 and 8-5 describe the markings and types of finishes applied to the major parts.

Figure 8-9. Overall view of the Model 1953 Mauser rifle.

Serbian and Yugoslav Mausers

Table 8-3
Technical Data, Post-World War II Yugoslav Rifles

RIFLE	Yugoslav Mauser Rifle M1948, M48A, M48B, M48BO	Yugoslav Mauser Rifle M1953	Yugoslav Mauser Rifle M24/47	Yugoslav Mauser Rifle M48/52-Č	Yugoslav Mauser Rifle M98[n] (Mod. 98) with long barrel	Yugoslav Mauser Rifle M98(n) (M.98)
Calibre	7.92 x 57 mm					
Rifling	concentric, 4 grooves, 0.15 mm deep and 4.3 mm wide; 1 turn in 240 mm, right hand (pitch of 5° 54')		concentric, 4 grooves, 0.13 - 0.18 mm deep and 4.2 - 4.4 mm wide; 1 turn in 240 mm, right hand (pitch of 5° 54')		concentric, 4 grooves 0.125 mm deep and 4.4 mm wide; 1 turn in 240 mm, right hand (pitch of 5° 54')	concentric, 4 grooves, 0.15 mm deep and 4.4 mm wide; 1 turn in 240 mm, right hand (pitch of 5° 54')
Magazine	internal staggered - column box, capacity 5 rounds					
Loading system	charger, or single rounds					

Serbian and Yugoslav Mausers

Length overall	mm	1095	1000	1100	1095	1250	1110-1,115
	inch	43.1	39.4	43.3	43.1	49.2	43.7-43.9
Barrel length	mm	590.2	580	590.2	589	740	600
	inch	23.24	22.8	23.24	23.2	29.1	23.6
Weight	gm	4100	3800	3,680-4,050	3890 ± 200	4150	3800
	lbs	9	8.38	8.1-8.3	8.57-9.02	9.1	8.38
Sights		(front) protected barleycorn; (rear) a tangent-leaf sight graduated from 200-2000 meters (218.5-2,187.2 yards)	(front) protected barleycorn; (rear) a tangent-leaf sight graduated from 200-2000 meters (218.5-2,187.2 yards)	(front) unprotected barleycorn; (rear) a tangent-leaf sight graduated from 200-2000 meters (218.5-2,187.2 yards)	(front) unprotected or protected barleycorn; (rear) a tangent-leaf sight graduated from 300-2000 meters (328-2,187.2 yards)	(front) unprotected barleycorn; (rear) a tangent-leaf sight graduated from 100-2000 meters (109.4-2,187.2 yards)	(front) open or protected barleycorn; (rear) a tangent leaf sight graduated from 100-2000 meters (109.4-2,187.2 yards)
Action	inch	Intermediate-ring Mauser, short action. Receiver ring dia., 35.8 mm (1.41 in); receiver length, 215.9 mm (8.5 in); screw spacing, 193.5 mm (7.62 in); bolt body length 155.3 mm (6.115 in); magazine length, 820.9 mm (3.232 in).			Large-ring M98 Mauser action. Receiver ring dia., 35.8 mm (1.41 in); receiver length, 222.25 mm (8.75 in); screw spacing, 199 mm (7.835 in); bolt body length 161.3 mm (6.35 in); magazine length, 842 mm (3.315 in).		

Serbian and Yugoslav Mausers

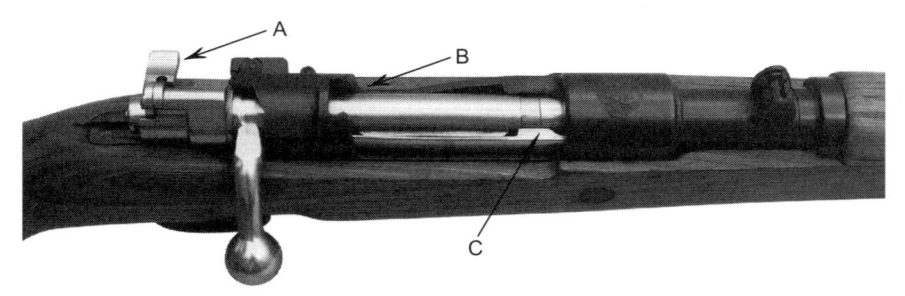

Figure 8-10. Model 1953 receiver and rear sight.

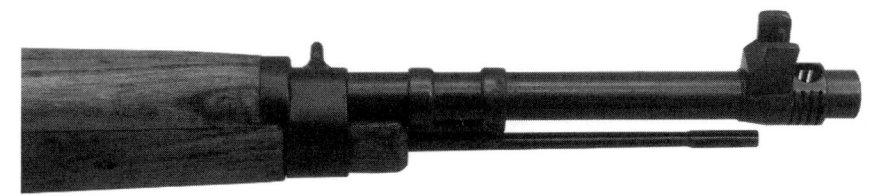

Figure 8-11. Model 1953 front sight.

CROSS SECTION

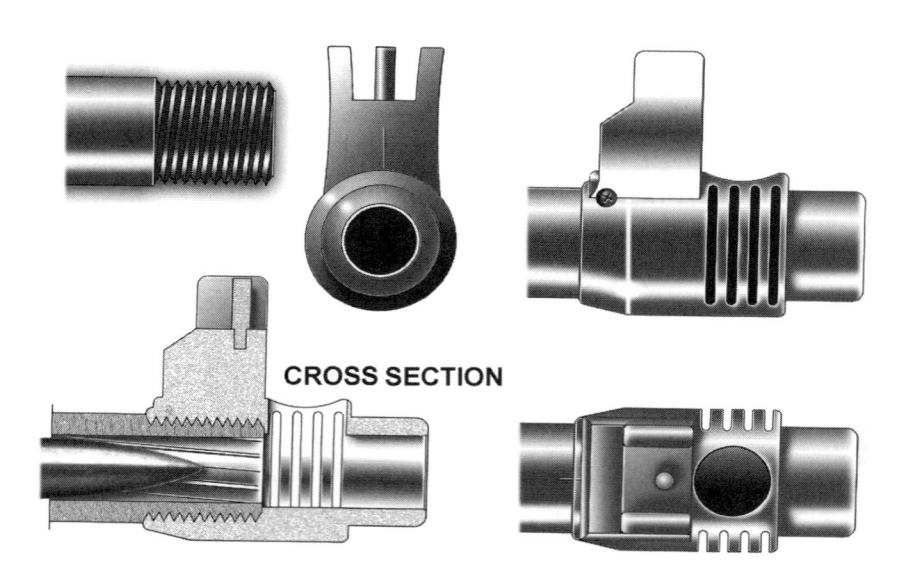

Figure 8-12. Cross section view of the Model 1953 front sight and recoil compensator.

Serbian and Yugoslav Mausers

Ses Also Table C-1, C-2, C-3, C-4, C-5, C-6, C-7, C-8	Coat of Arms	Pattern Mark	Repair Workshop's Marks	Country's Name	Year of Production	Ruler's Monogram	Full # Number	Last Three Digits of Serial Number	Smokeless Powder Proof Mark	Inspection Mark	Acceptance Mark	Calibre
Barrel:								x y	x y	x y	x y	x y
Receiver Ring:												
Top	x y											
Left			y	x					x y	x y		
Right							x y					
Receiver												
Left		x y										
Right												
Recoil Lug										x y		
Bolt Assembly												
Bolt Body												
Bolt Handle Base										x y		
Bolt Handle Neck							x y					
Bolt Handle Ball									x y			
Cocking Piece										x y		

Table 8-4
Markings,
Mauser Rifle M98/48, 7.9 mm

Serbian and Yugoslav Mausers

Table 8-4, cont.
Markings,
Mauser Rifle M98/48, 7.9 mm

Ses Also Table C-1, C-2, C-3, C-4, C-5, C-6, C-7, C-8	Coat of Arms	Pattern Mark	Repair Workshop's Marks	Country's Name	Year of Production	Ruler's Monogram	Full # Number	Last Three Digits of Serial Number	Smokeless Powder Proof Mark	Inspection Mark	Acceptance Mark	Calibre
Bolt Sleeve with Gas Shield								x y				
Safety Lever								x y				
Extractor												
Extractor Collar												
Ejector												
Bolt Stop										x y		
Trigger Assembly												
Trigger												
Sear-fork										x y		
Trigger Guard Plate									x y			
Magazine Floor Plate							x y			X y		
Stock				x y					x y			
Butt Plate							x y					
Front Band, Lower Band										X y		

x = Type I (for Military Use)
y = Type II (for Export)

Serbian and Yugoslav Mausers

<table>
<tr><td colspan="2" align="center">Table 8-5
Finishes,
M1948, M1948A, M1948B, M24/47,
M98/48, M24/52-Č, M53 Rifle, and M1969 Sniper Rifle</td></tr>
<tr><th>Part</th><th>Finish</th></tr>
<tr><td>All wood parts</td><td>Sanded smooth and oiled</td></tr>
<tr><td>Barrel</td><td>Blued, used a hot salt bath method after polishing and degreasing</td></tr>
<tr><td>Receiver</td><td>Blued, used a hot salt bath method after polishing and degreasing</td></tr>
<tr><td>Front Band, Bayonet Mount, Lower Band</td><td>Blued, used a hot salt bath method after polishing and degreasing</td></tr>
<tr><td>Bolt Disassembly Disc</td><td>Finished in-the-white</td></tr>
<tr><td>Butt Plate</td><td>Finished in-the-white (1)</td></tr>
<tr><td>Recoil Lug</td><td>Blued by heating</td></tr>
<tr><td>Front Sight Blade</td><td>Blued by heating</td></tr>
<tr><td>Front Sight Hood</td><td>Blued by heating</td></tr>
<tr><td>Rear Sight Base</td><td>Blued, used a hot salt bath method after polishing and degreasing</td></tr>
<tr><td>Sight Leaf</td><td>Blued (2)</td></tr>
<tr><td>Slide</td><td>Blued</td></tr>
<tr><td>Slide Catch</td><td>Blued</td></tr>
<tr><td>Screws</td><td>Finished with a black coating</td></tr>
<tr><td>Bolt Parts</td><td>Finished in-the-white (3)</td></tr>
<tr><td>Bolt Stop</td><td>Blued by heating</td></tr>
<tr><td>Bolt Sleeve Lock Plunger</td><td>Blued by heating</td></tr>
<tr><td>Trigger Guard and Floor Plate</td><td>Blued, used a hot salt bath method after polishing and degreasing</td></tr>
</table>

Serbian and Yugoslav Mausers

Table 8-5, cont. Finishes, M1948, M1948A, M1948B, M24/47, M98/48, M24/52-Č, M53 Rifle, and M1969 Sniper Rifle	
Part	Finish
Trigger	Blued
Sear	Finished in-the-white
Follower	Blued
Front and Rear Trigger Guard Screws, Front Lock Screw, Rear Lock Screw	Blued
Trigger Guard Screw Tube	Finished in-the-white
Magazine Spring	Blued spring steel
Floor Plate Catch	Blued by heating
1 Some specimens of the M24/47, M98/48 & M24/52-Č, were blued after being repaired. 2 Some specimens of the M24/47 were polished in-the-white. 3 Some repaired specimens of the M98/48 had all bolt parts blued.	

Export Model Yugoslavian Mausers

The M48 rifle remained in production longer than any other domestic Mauser model. It was also the first domestically produced firearm sold by Yugoslavia to a foreign country. In 1953, Burma contracted for 10,000 Model 1948A rifles and 7,000,000 7.92 x 57 mm rounds of ammunition. During the period 1957 to 1960, Yugoslavia also sold the Model 1948A model to Syria, Indonesia, Egypt, and Iraq, see Figures 8-13, 8-14, 8-15, and 8-16.

Yugoslavia supported the so-called "National Liberation movements" of the period with shipments of Model 1948 rifles. Many of these were manufactured without markings and known as the "BO" pieces *(bez oznake* = unmarked).

One of more important 20th-century international political scandals was tied to the delivery of Yugoslavian Model 1948 rifles to the Algerian revolutionary movement, ALN/FLN (armed wing of the National-Liberation Front). Algeria was then a French colony. A deputy of the FLN in Cairo, Lakhdar Brahimi (now a United Nations commissioner in Afghanistan as this was written), was married to Milica, daughter of a Yugoslav captain. Brahimi managed to divert, with Yugoslav help,

Serbian and Yugoslav Mausers

7.62 mm calibre *PAP* M59/66 & M59/66A1

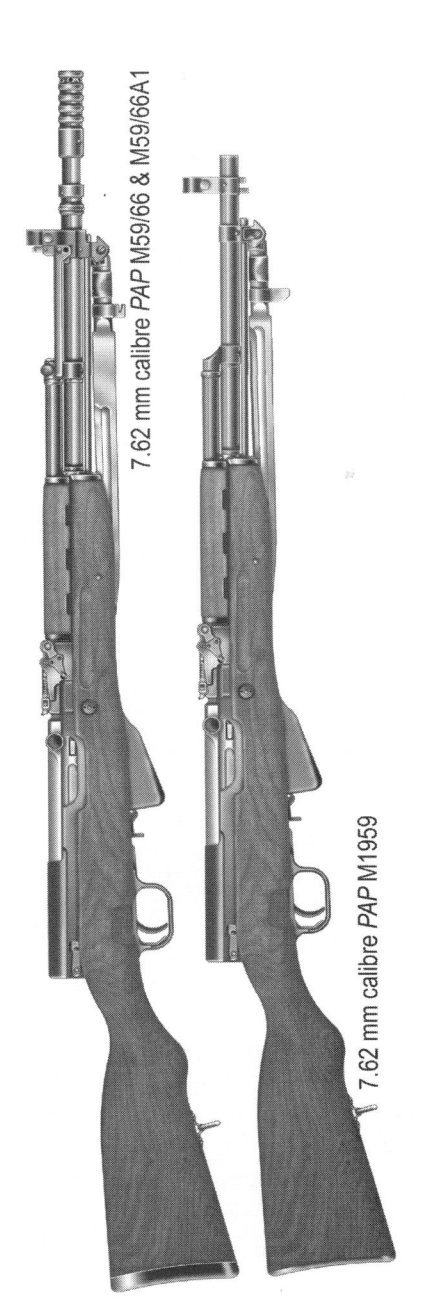

7.62 mm calibre *PAP* M1959

Figure 8-13a. 7.62 x 39 mm calibre *PAP* M59/66 and M59/66A1.

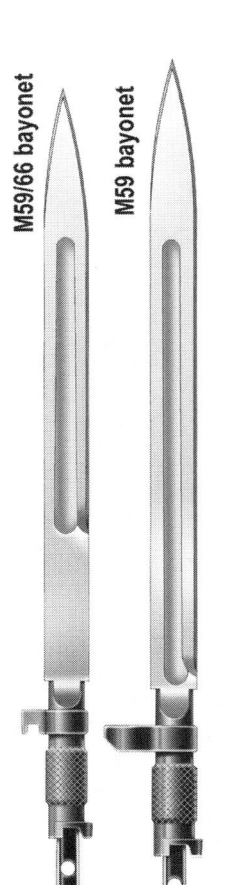

M59/66 bayonet

M59 bayonet

Figure 8-13b. The folding bayonet for the M59/96 and M59.

Serbian and Yugoslav Mausers

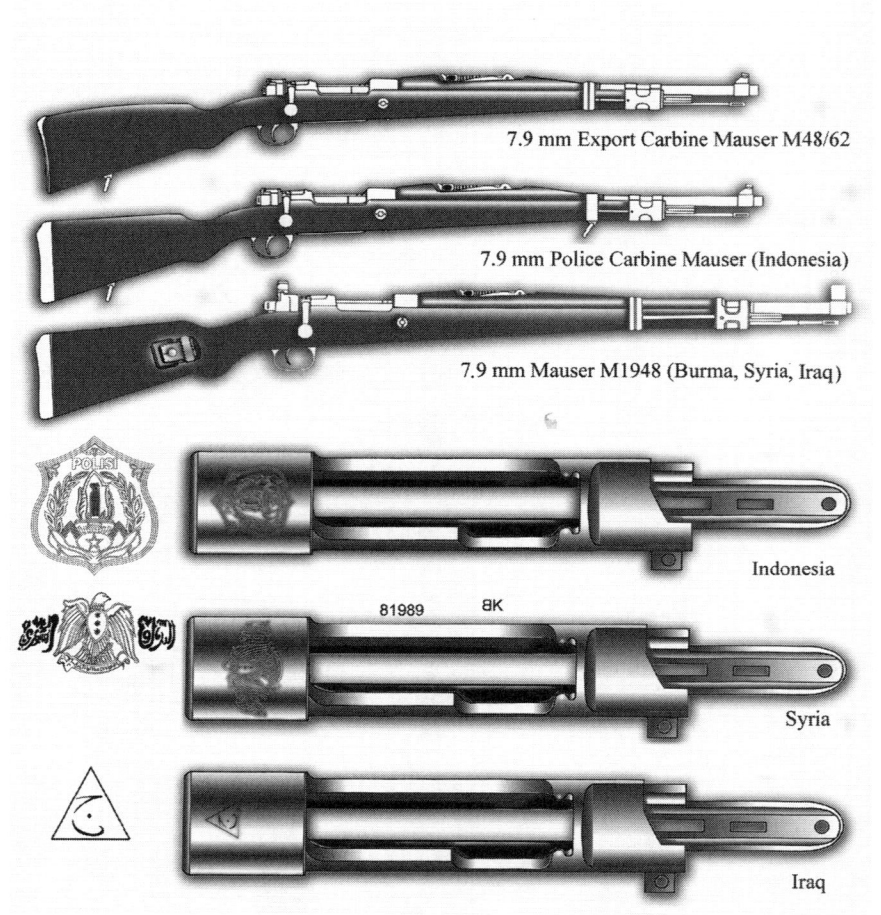

Figure 8-14. Export variations of the Model 1948 7.92 x 57 mm Mauser rifle and carbines.

50,000 Model 1948B 7.92 x 57 mm rifles from the contingent intended for Iraq to the ALN.

On 18 January 1958, the French Navy stopped the ship, *Ljubljana*, which was allegedly transporting 148 metric tons of Czechoslovakian arms and munitions for ALN combatants (which were training in camps in Morocco). But the majority of the shipment, in fact, consisted of Yugoslav Model 1948 7.92 mm rifles and the result was a serious political dispute with France.

Serbian and Yugoslav Mausers

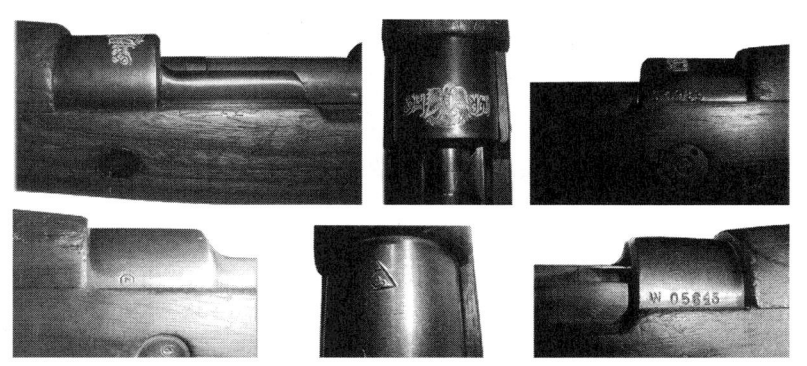

Figure 8-15. Receiver markings on Syrian (top) and Iraqi (bottom) Mauser rifles made in Yugoslavia.

Figure 8-16. Model 1948 Mauser carbine made for the Indonesian police.

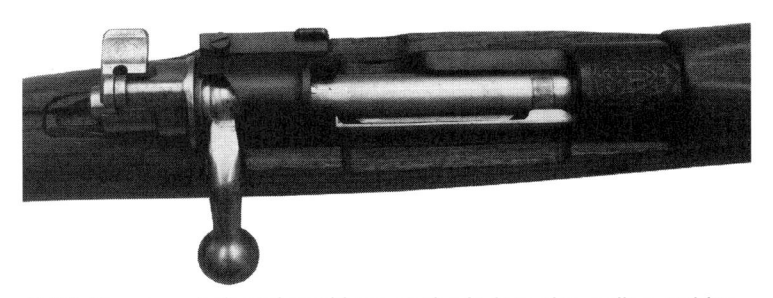

Figure 8-17. Receiver, bolt and markings on the Indonesian police carbine.

In 1962, the types of Mauser rifles then in Yugoslav arsenals are shown in Table 8-6.

During the Cold War and after, Yugoslavian Mauser rifles were exported to Third World countries in great quantity. The central storage depot at Mostar in Bosnia and Hercegovina also distributed Mausers to museums, movie companies, and sold many to arms dealers in Canada and the United States. A 7.92 x 57 mm Mauser of any model was 700 dinars or $62.30 each in August 1970. Thirteen different models of Yugoslavian Mausers remained in storage that year, as shown in Table 8-7.

Serbian and Yugoslav Mausers

Table 8-6
1013-Rifle and Carbines, 1962 Inventory

Ordinal No.	Name	Model and Origin	Storage Number
114	Puska 7.9 mm M24 Rifle 7.9 mm M24	Yugoslav rifle 7.92 x 57 mm Mauser M24	1013-1023-0144
115	Puska 7.9 mm M24 (b) Rifle M24(b) 7.92 mm	Belgian Mauser 7.92 x 57 mm M24	1013-1023-0145
116	Puska 7.9 mm M24 (c) Rifle M24(c) 7.9 mm	Czech Mauser vz.24 7.92 x 57 mm - the Rumanian variant ZB md.24, German "Gewehr" 24(t)	1013-1023-0146
117	Puska 7.9 mm M24/47 Rifle M24/47 7.9 mm	Yugoslav Mauser M24 7.92 x 57 mm assembled after WW II	1013-1023-0139
118	Puska 7.9 mm M24/52 (c) Rifle M24/52 (c) 7.9 mm	Repaired vz.24, "Gewehr" 24(t) and "ZB" md.24 7.92 x 57 mm	1013-1023-0140
119	Puska 7.9 mm M48 Rifle M48 7.9 mm	Yugoslav Mauser M48 7.92 x 57 mm	1013-1023-0135
120	Puska 7.9 mm M48A Rifle M48A 7.9 mm	Yugoslav Mauser M48A 7.92 mm	1013-1023-0137
121	Puska 7.9 mm M48B Rifle M48B 7.9 mm	Yugoslav Mauser rifle M48B 7.92 mm	1013-1023-0138
122	Puska 7.9 mm M48 bez oznake (BO) Rifle M48 (BO) (no markings) 7.9 mm	Yugoslav rifle M48 7.92 x 57 mm for export	1013-1023-0136
123	Puska 7.9 mm M98 (n) Rifle M98 (n) 7.9 mm	German Carbine Mauser K98k 7.92 mm, repaired prior to 1950 (new model markings M.98 or Mod.98)	1013-1023-0147

Serbian and Yugoslav Mausers

Ordinal No.	Name	Model and Origin	Storage Number
	Table 8-6, cont.		
	1013-Rifle and Carbines, 1962 Inventory		
124	Puska M98 (n), sa dugom cevi 7.9 mm Rifle M98 with long barrel 7.9 mm	German Mauser M1898 7.9 x 57 mm, repaired prior to 1950 (new model markings M.98 or Mod.98)	1013-1023-0148
125	Puska 7.9 mm M98/48 (n) Rifle M98/48 (n) 7.9 mm	Repaired K98k 7.92 mm, repaired after 1950 (new model markings M.98/48)	1013-1023-0141
126	Puska 7.9 mm M98/48 (n), sa dugom cevi Rifle M98/48 7.9 mm with long barrel	German Mauser M1898 7.92 mm, repaired after 1950 (new model markings M.98/48)	1013-1023-0143
127	Puska 7.9 mm M98/48 (n), bez oznake (BO) Rifle M48/98 (n) (BO) (no markings) 7.9 mm	Repaired rifle M1898 and carbine K98k 7.92 x 57 mm for export	1013-1023-0142

No.	Name	Model and Origin	Manufacturer's markings (code designations)*	YNA Storage Number	Control Number
		Table 8-7			
		1013—Rifles and Carbines, 1970 Inventory			
92	Puska 7.9 mm M48 Rifle M48 7.9 mm	Yugoslav Mauser M48 7.92 mm	1362	1013-1023-0135	0
93	Puska 7.9 mm M48 bez oznake (BO) Rifle M48 no markings (BO) 7.9 mm	Yugoslav rifle M48 7.92 x 57 mm for export	1362	1013-1023-0136	1

Serbian and Yugoslav Mausers

| \multicolumn{5}{|c|}{Table 8-7, cont.
1013—Rifles and Carbines, 1970 Inventory} |

No.	Name	Model and Origin	Manufacturer's markings (code designations)*	YNA Storage Number	Control Number
94	Puska 7.9 mm M48A Rifle M48A 7.9 mm	Yugoslav Mauser M48A 7.92 x 57 mm	1362	1013-1023-0137	3
95	Puska 7.9 mm M48B Rifle M48B 7.9 mm	Yugoslav Mauser rifle M48B 7.92 x 57 mm (model markings – M48A)	1362	1013-1023-0138	5
96	Puska 7.9 mm M24 Rifle 7.9 mm M24	Yugoslav rifle 7.92 x 57 mm Mauser M24	4387	1013-1023-0144	0
97	Puska 7.9 mm M24 (c) Rifle M24(c) 7.9 mm	Czech Mauser vz.24 7.9 mm-the Rumanian variant ZB md.24, German "Gewehr" 24(t)	4387	1013-1023-0146	4
98	Puska 7.9 mm M24/47 Rifle M24/47 7.9 mm	Yugoslav Mauser M24 7.92 x 57 mm assembled after WW II	4387	1013-1023-0139	7
99	Puska 7.9 mm M24/52 (c) Rifle M24/52 (c) 7.9 mm	Repaired vz.24, "Gewehr" 24(t) and "ZB" md.24	4387	1013-1023-0140	3

Serbian and Yugoslav Mausers

	Table 8-7, cont. 1013—Rifles and Carbines, 1970 Inventory				
No.	Name	Model and Origin	Manufacturer's markings (code designations)*	YNA Storage Number	Control Number
100	Puska 7.9 mm M98/48 (n) Rifle M98/48 (n) 7.9 mm	German carbine Mauser K98k 7.92 x 57 mm, repaired after 1950 (model markings M.98/48)	4387	1013-1023-0141	5
101	Puska 7.9 mm M98/48 (n), bez oznake Rifle M48/98 (n) – no markings (BO) 7.9 mm	Repaired rifle M1898 and carbine K98k 7.92 x 57 mm for export (without model markings)	4387	1013-1023-0142	7
102	Puska 7.9 mm M98/48 (n) sa dugom cevi Rifle M98/48 7.9 mm with long barrel	German rifle Mauser M1898 7.92 mm, repaired after 1950 (model markings M.98/48)	4387	1013-1023-0143	9
103	Puska 7.9 mm M98 (n) Rifle M98 (n) 7.9 mm	German carbine Mauser K98k 7.92 x 57 mm, repaired prior to 1950 (model markings M.98 or Mod.98)	4387	1013-1023-0147	6
104	Puska 7.9 mm M69, Snajper Sniper M69 7.9 mm	Yugoslav sniper Mauser M1969 7.92 x 57 mm	1362	1013-1140-7088	0

* See Table C-1 through C-8 in Appendix C.

CHAPTER 9
SNIPER RIFLES

As one of countries aligned with the Soviet Union in 1947, Yugoslavia received a loan from the USSR for the purchase of arms. The loan amounted to $78,000,000 with the condition that it be paid back within ten years at 2 percent interest.

THE MAUSER 7.92 MM M48/52, M1969 AND M1993 SNIPER RIFLES

In part this loan financed Moscow's delivery of 4,580 7.62 mm Mosin-Nagant M1891/30-M31 sniper rifles with PU 3x telescopic sight with a range up to 1,300 meters (1,421.6 yards). The PU 3x telescopic sight was developed in 1937 and was based on a telescopic sight designed and manufactured by the German firm of Busch & Ratenow.

The Yugoslavian army soon started to develop its own telescopic sights. Its first development and production facility was an improvised workshop, "Zrak," located at Knezevac near Belgrade. A year earlier (1948), the Zvijezda aircraft instrument factory, located outside Sarajevo in Bosnia and Hercegovina, embarked on an effort to produce optical sights; however, it was soon decided to concentrate all optical production facilities. Consequently, in mid-1951, the government moved the Knezevac facility to the Zvijezda plant and the factory as a whole was renamed "Zrak." The plant was not immediately operational as it took until 1956 to develop the technology and machinery necessary for full production of optical sights.

Yugoslavia already had the ability to produce sniper rifles. As the Zastava or "Red Flag Enterprise" factory had proven quite capable in the manufacture of the M48A rifle, in 1953 the Army's Infantry Department directed it to begin production of sniper rifles. Initially, the effort was limited to selecting 4,618 rifles and equipping them with the M52 optical sight which was a Yugoslav copy of the Russian PU sight, see Figures 9-1, 9-2 and 9-3. The early series of the M52 came from the Tovarna opticnih sredstev (TOS or Optical Sights Factory) in Ljubljana, Slovenia, since the Zrak facility was still in the process of being built. Tovarna opticnih sredstev, Ljubljana, is now known as Iskra Kibernetika, Tovarna Vega p.o., Ljubljana, Slovenia.

Though the country had a capable weapons industry, it was not until the 1970s that the Yugoslavian army requested its own specially designed sniper rifle. Kragujevac's Zastava factory received the order to design the rifle, and the design team was led by Milutin Milojevich and Milosh

Serbian and Yugoslav Mausers

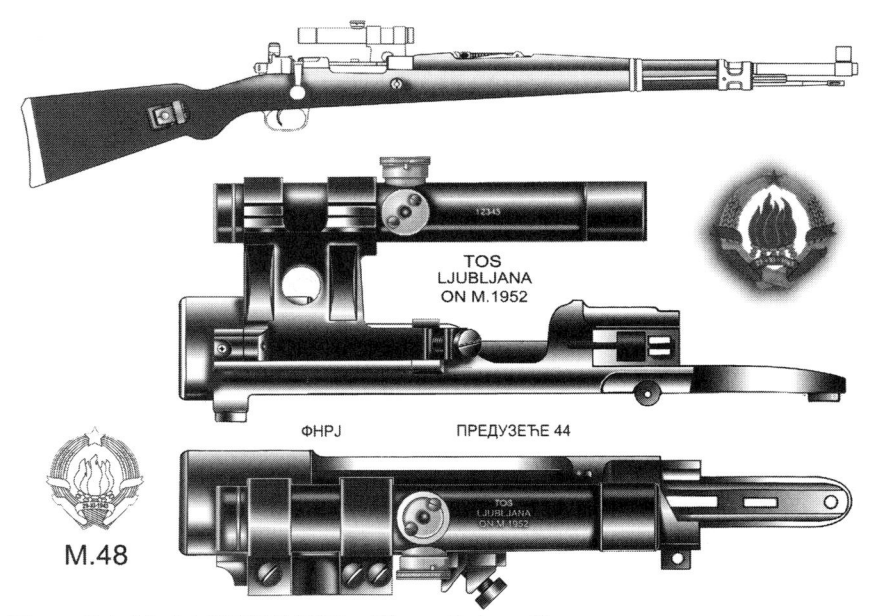

Figure 9-1. Model 1948/52 7.92 x 57 mm Sniper rifle.

Ostojich who developed the five-shot Model 1969 in 7.92 mm calibre, see Figures 9-4 and 9-5. Sarajevo's Zrak facility got the contract to develop the Model 1969's telescopic sight (Figure 9-6) and Uzice's First Partisan factory gained the contract to produce the rifle's ammunition.

The Model 1969 7.92 x 57 mm sniper rifle was a true domestically designed and produced firearm. It was developed from Zastava's experiences in military and sporting arms production, particularly the LK-70 sporting carbine, refer to Figure 9-5 and see also Figure 11-1.

It featured the Mauser Model 1898's turn bolt. Because it had a longer travel than that of the Yugoslav Model 1924 and M1948 rifles (6 mm or 0.236 inch), it cycled somewhat slower at 5 to 8 rounds per minute. But Zastava's engineers were concerned with easier bolt handling. The turned-down bolt handle design (Figures 9-5A and 9-5B) taken from the carbine version made it easier to cock and, at the same time, less likely to catch in dense foliage. The use of an optical sight in turn demanded a new butterfly-shaped safety catch. Other features included a trigger mechanism with a single trigger set, a direct result of the factory's experience with development of sporting carbines.

Serbian and Yugoslav Mausers

Figure 9-3. TOS (Optical Sights Factory, Ljubljana, Slovenia) ON-M52 telescopic sight and mount for the Model 1948/52 Sniper rifle.

In some respects the trigger mechanism resembled the classic Mauser military mechanism but it had another notch that facilitated continuous and safe firing. The optical sight's placement directly above the receiver eliminated the possibility of loading from a clip. The special ammunition had to be loaded round by round. This necessitated moving the magazine plate catch from the inner side of trigger guard to the outside of the trigger guard, thus leaving more free space around the trigger. The new trigger mechanism also provided a smoother trigger squeeze, and the extra space gained by moving the magazine catch made it easier to shoot when wearing winter gloves. On the other hand, the magazine design was more complex technologically and utilized mechanically treated, forged pieces.

The Model 1969 7.92 x 57 mm sniper rifle used the ON-2 M1969 optical sight produced by the Zrak facility, see Figure 9-6. Yugoslavia kept the Model 1969 sniper rifle in service for almost 30 years until replaced by the Model 1991/1993 7.92 x 57 mm repeating sniper rifle, designed by Zastava's engineers led by Marinko Petrovich.

Figure 9-2. Overall view of the Model 1948/52 Sniper rifle.

Serbian and Yugoslav Mausers

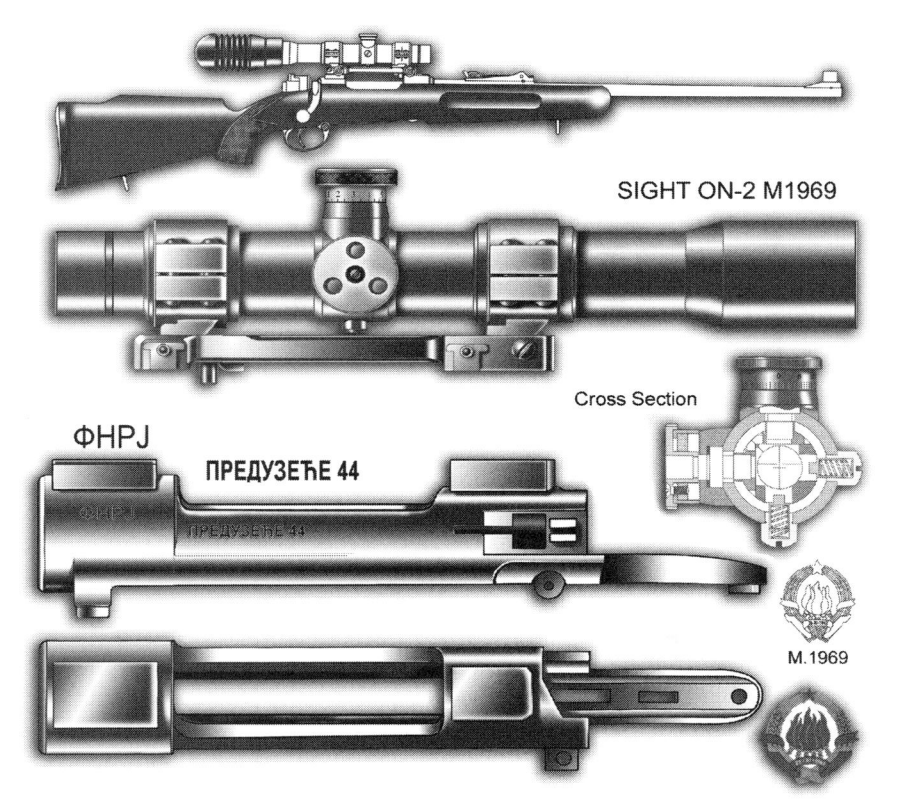

Figure 9-4. Detailed view of the Model 1969 Sniper rifle showing the receiver markings and internal construction of the windage and elevation adjusting mechanisms.

The export version of the Model 1991/1993 was rechambered to fire the NATO's 7.62 x 54 mm cartridge.

THE LONG-RANGE SNIPER RIFLE MODEL 1993 12.7 MM "BLACK ARROW"

Serbian weapons specialists carefully analyzed the capabilities of the new Anti Material Rifles (AMR), during the armed conflicts of the 1980s and 1990s. A change in government resulted in Kragujevac's arms factory being renamed the Zastava Namenski Proizvodi. Its engineers found the performance of the various AMRs to be impressive. Led by Dragoljub Grujovich and Radmilo Lepojevich, the Zastava factory set out to develop a similar weapon, see Figure 9-7.

Serbian and Yugoslav Mausers

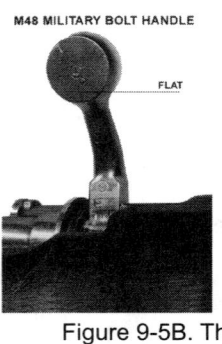

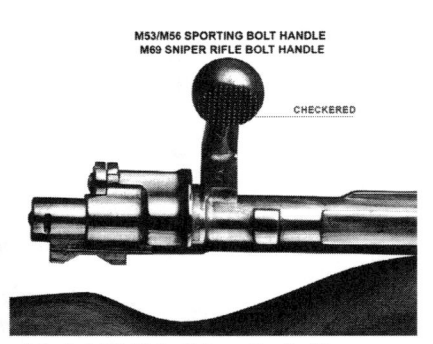

Figure 9-5B. The Model 1948 bolt handle (left) compared to the Model 1969 Sniper rifle and Model 1953/56 sporting rifle bolt handles (right).

The engineers merged the concept of the AMR repeating rifle with the factory's long experience with the Mauser bolt-action system. Dragoljub Grujovich selected the Model 1898 bolt, retaining its overall configuration but proportionally enlarging it to handle the 12.7 x 99 mm (.50 Browning Machine Gun) and 12.7 x 107 mm (12.7 mm Soviet Machine Gun) cartridges. The resulting rifle was designated the Model 1993, see Figure 9-8.

To prevent the bolt from bending under the chamber pressures produced by the 12.7 mm cartridge, it incorporated three massive locking lugs which distributed the pressure symmetrically. The 12.7 mm cartridge had a maximum pressure of 380 megapascals (Mpa) or 8 lb/ft^2 while the 7.92 x 57 mm cartridge produced 310 Mpa (6.5 lb/ft^2). In case of a ruptured case, the escaping gases were channeled out of the breech through two gas ports on the left side of the bolt body. This design was similar to that used in the Model 1898 Mauser and the Yugoslavian Models of 1924 and 1948.

The upper left side of the bolt's body featured a longitudinal rib that helped guide its travel by bearing in a slot on the receiver bridge. The strong steel extractor typical of Mauser weapons

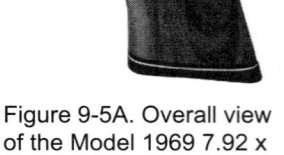

Figure 9-5A. Overall view of the Model 1969 7.92 x 57 mm Sniper rifle.

Serbian and Yugoslav Mausers

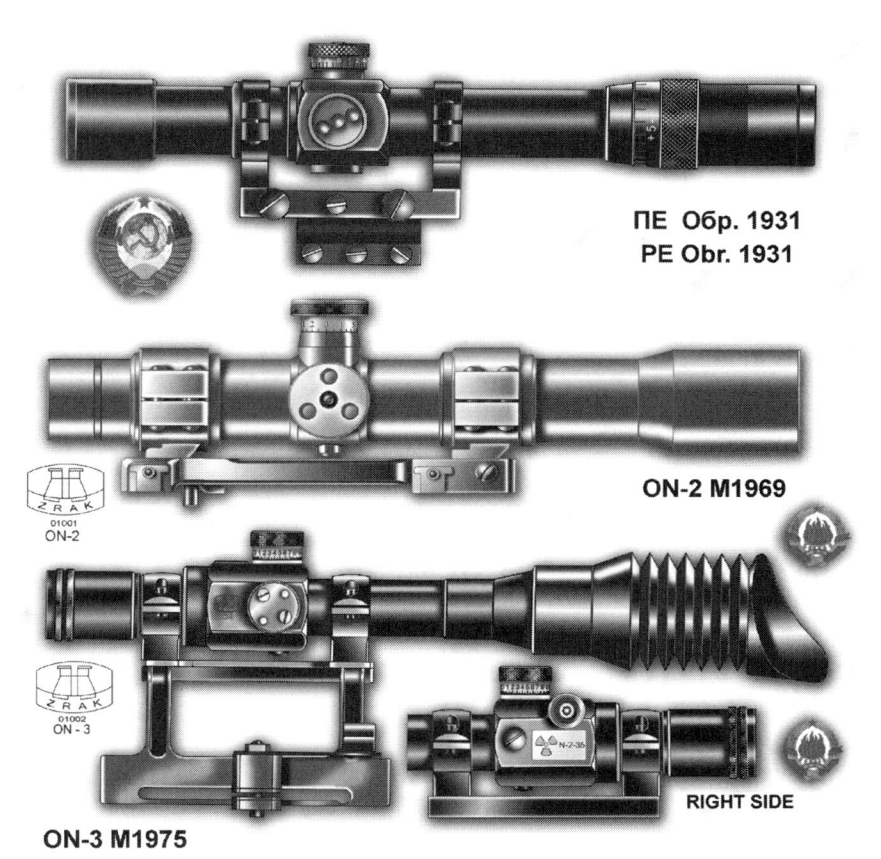

Figure 9-6. Detail views: (above) USSR PE Pattern 1931, (middle) Yugoslavian ON-2 M69, and (below) Yugoslavian ON-3 M76.

was placed on the bolt body's right side and tied to it by the extractor collar. The bolt handle was flat and turned down. A 90-degree turn cocked the firing pin.

One of the Model 1993 sniper rifle's unusual features was its simplified bolt sleeve stop and cocking piece that resulted from the new safety catch design. It consisted of a lever-type safety catch that blocked the trigger regardless of the position of the firing pin. This safety catch was simpler and safer than the one designed for the Model 1959 SKS Carbine. A heavy barrel made of chromium-nickel-vanadium steel considerably reduced longitudinal oscillations, supported precision on firing, and guaranteed ballistic consistency in excess of 5,000 rounds. Barrel

Serbian and Yugoslav Mausers

Figure 9-7. The designers of the Yugoslav Model 1993 .50 caliber Browning "Black Arrow" Sniper rifle: (left to right) Dragan Lishanin, Dragisha Antonijevich, Dragoljub Grujovich and Radmilo Lepojevich.

ribbing provided more cooling area. The rifle was loaded with a detachable 5-shot box magazine, and the magazine catch was located on the receiver's right side.

Engineers reduced the bullet's recoil energy of about 128 Joules (95 ft lb) a recoil compensator on the muzzle. They also installed a spring-loaded buffer in the buttstock and added a rubber butt plate. The compensator, buffer and butt pad lowered recoil energy to 50 to 55 Joules (37 to 40.7 ft lb), equivalent to that of a .300 Winchester Magnum sporting rifle.

The Model 1993's design included a bipod and a folding carry handle for transportation. In firing tests using the Model 1994 telescopic sight,

Figure 9-8. Overall view of the Model 1993 .50-calibre Browning "Black Arrow" Sniper rifle.

Serbian and Yugoslav Mausers

the 8 x 56 Zrak-Teslic scope, and match ammunition, the new sniper rifle achieved one minute of angle degree (1 MOA) accuracy at 1,500 meters (1,604.4 yards).

See Table 9-1 for specifications for Yugoslav Sniper Rifles based on the Mauser action and Table 9-2 for all markings.

RIFLE		Yugoslav Sniper Rifle M1948/52	Yugoslav Sniper Rifle M1969	Yugoslav Sniper Rifle M1993	Yugoslav Sniper Rifle M1993 "Black Arrow"
Calibre	mm	7.92	7.92	7.92	11.43 (.50 BMG)
Rifling		concentric, 4 grooves, 0.15 mm deep and 4.3 mm wide; 1 turn in 240 mm, right hand (pitch of 5° 54')	concentric, 4 grooves, 0.155 mm deep and 4.2 mm wide; 1 turn in 240 mm, right hand	concentric, 4 grooves, 0.155 mm deep and 4.2 mm wide; 1 turn in 240 mm, right hand	concentric, 8 grooves, 0.18 mm deep and 2.8 mm wide; 1 turn in 381 mm, right hand
Magazine		internal staggered - column box, capacity 5 rounds	internal staggered - column box, capacity 5 rounds	detachable external staggered - column box, capacity 5 or 10 rounds	detachable external staggered - column box, capacity 5 rounds
Loading system		charger, or single rounds	single rounds	single rounds	single rounds
Length overall	inch	43.1	44.5	50	59.5
	mm	1095	1130	1270	1510

Serbian and Yugoslav Mausers

	Table 9-1, cont. Technical Data Yugoslav Sniper Rifles				
RIFLE		Yugoslav Sniper Rifle M1948/52	Yugoslav Sniper Rifle M1969	Yugoslav Sniper Rifle M1993	Yugoslav Sniper Rifle M1993 "Black Arrow"
Barrel length	inch	23.24	23.62	27.5	33
	mm	590.2	600	700	840
Weight	lbs	9	10.50	14.30	32
	g	4100	4800	6500	14450
Optical Sight		ON-M52 TOS Ljubljana	ON-2 M1969 Zrak-Sarajevo	3-9x42 or 6x42	M94 Zrak-Teslic, 8x56
Action		Intermediate-ring Mauser, short action. Receiver ring dia., 35.8 mm (1.41 in); receiver length, 215.9 mm (8.5 in); screw spacing, 193.5 mm (7.62 in); bolt body length 155.3 mm (6.115 in); magazine length, 820.9 mm (3.232 in).	Large-ring M98 Mauser action. Receiver ring dia. 35.8 mm (1.41 in); receiver length, 222.25 mm (8.75 in); screw spacing, 199 mm (7.835 in), bolt body length 161.3 mm (6.35 in), Magazine length, 8.42 mm (3.315 in)	Large-ring M98 Mauser action. Receiver ring dia. 35.8 mm (1.41 in); receiver length, 222.25 mm (8.75 in); screw spacing, 199 mm (7.835 in), bolt body length 161.3 mm (6.35 in), Magazine length, 8.42 mm (3.315 in)	

Serbian and Yugoslav Mausers

Table 9-2
Markings,
Mauser Rifle M1948, M1948A, M1948B, M24/52-Č 7.9 mm

	See Also Table C-1, C-2, C-3, C-4, C-5, C-6, C-7, C-8	Coat of Arms and Pattern Mark	Manufacturer's Markings	Country's Name	Year of Production	Ruler's Monogram	Full Serial Number	Last Three Digits of Serial Number	Smokeless Powder Proof Mark	Inspection Mark	Acceptance Mark	Calibre
Barrel								x	x	x	x	
Receiver Ring												
Top		x										
Left			x						x	x		
Right						x						
Receiver												
Left			x									
Right												
Recoil Lug										x		
Bolt Assembly												
Bolt Body												
Bolt Handle Base										x		
Bolt Handle Neck							x			x		
Bolt Handle Ball									x			
Firing Pin												
Cocking Piece										x		
Bolt Sleeve with Gas Shield									x	x		
Bolt Sleeve Stop												
Safety Lever									x	x		

Serbian and Yugoslav Mausers

Table 9-2, cont. Markings, Mauser Rifle M1948, M1948A, M1948B, M24/52-Č 7.9 mm											
See Also Table C-1, C-2, C-3, C-4, C-5, C-6, C-7, C-8	Coat of Arms and Pattern Mark	Manufacturer's Markings	Country's Name	Year of Production	Ruler's Monogram	Full Serial Number	Last Three Digits of Serial Number	Smokeless Powder Proof Mark	Inspection Mark	Acceptance Mark	Calibre
Extractor											
Extractor Collar											x
Ejector											
Bolt Stop									x		
Trigger Assembly											
Trigger									x		
Sear-fork									x		
Trigger Guard Plate						x			x		
Magazine Floor Plate						x			x		
Stock						x			x	x	
Butt Plate						x			x		
Front Band, Lower Band									x		

CHAPTER 10
SMALL-CALIBRE RIFLES

In 1954, Kragujevac's engineers began development of Mauser-based small-calibre rifles. In 1956, series production of the successful standard model 5.6 mm (.22-calibre long rifle) Model 48/52 and Model 1956 small-calibre rifles began, see Figure 10-1. These rifles were equipped with sights graduated for distances ranging from 25 to 200 meters (27.3 to 218.7 yards), see Figure 10-2. During the first year, 1956, only thirty-nine single-shot small-calibre Model 1956 rifles were manufactured. All were sent to the Shooters Association of Yugoslavia (SSJ). The rifle was marked on the receiver ring, "KRAGUJEVAC/M56" beneath the factory emblem, "CZ" inside a triangle for Crvena Zastava, or Red Flag, see Figure 10-3. The calibre marking, ".22 Long Rifle" was stamped on the left side of the barrel, just ahead of the receiver.

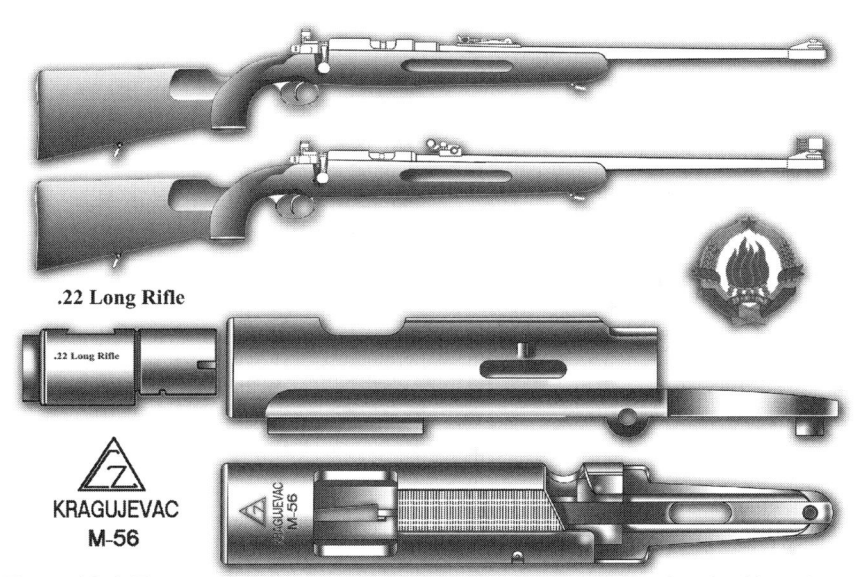

Figure 10-1. Two types of small-calibre sporting rifles developed at the Kragujevac factory: above, the Model 48/52 and below, the Model 1956 rifles.

The Model 1956 was replaced the following year with the new small-calibre, repeating Model 1957 rifle. This rifle was equipped with a micrometer rear sight and designed as a hunting rifle. A variation of the

Serbian and Yugoslav Mausers

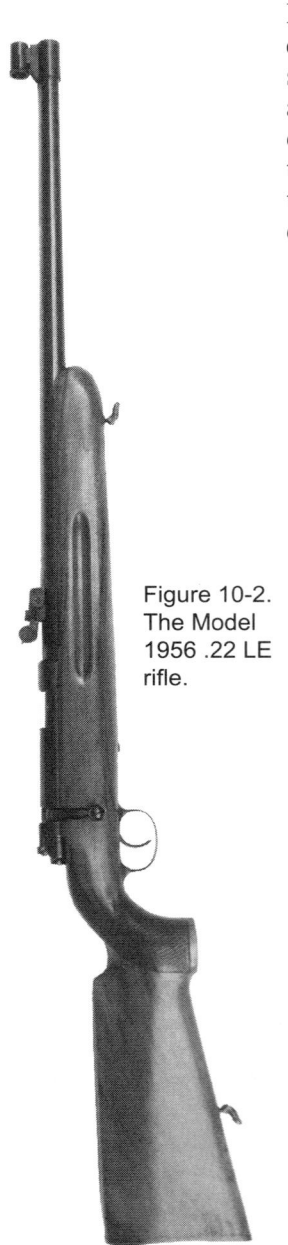

Model 1957 was the very accurate Miomanovich-designed Model 1958 MK rifle with diopter rear-sight, hooded front sight, and heavy, large-diameter barrel. In 1960, when series production of these models ended, the factory had manufactured a total of 12,099 small-calibre rifles, i.e., the M56, M57 and M58, all in .22 Long Rifle calibre, for the SSJ and commercial markets.

Figure 10-2. The Model 1956 .22 LE rifle.

Figure 10-3. Receiver ring marking on the Model 1956 .22-calibre rimfire rifle.

CHAPTER 11
SPORTING CARBINES

Because of its long experience in the production of military rifles, the Zastava factory was well positioned to move into the sporting-arms market. Development of sporting rifles began in 1953 with an order from the Forestry Department of Bosnia and Hercegovina for ninety-nine sporting carbines chambered for the 7.92 x 57 mm cartridge. Zastava took ninety-nine Model 1948A military rifles—which had just entered series production—and made minor modifications, primarily to the buttstocks, and sent them on to the Forestry Department at Sarajevo.

A second order from the Forestry Department followed a year later and the factory modified 129 more Model 1948A rifles into hunting carbines. Zastava's managers recognized the potential profits from the sale of sporting rifles and, in 1956, began series production of 1,017 Model 1956 repeating hunting carbines. Development of sporting weapons paralleled the process used by the factory for its military weapons. Engineers first analyzed sporting weapons produced by the leading sporting-arms manufacturers around the world while management conducted a market analysis to ascertain which models and calibres sold well. In 1962, management and engineers reached a decision on a design and the LK-62 carbine was placed into production. It was marketed in four variants. The LK-62/64 soon followed.

NOTE: The term "carbine" is used here as it was and is in common usage in Yugoslavia and Serbia to denote what would in many other countries be referred to as a rifle. Its usage here is not meant to imply that these models are "shorter" than the standard sporting rifle length.

The LK-62 presented unique challenges. Because of the trigger assembly's placement and the need for a magazine floor plate that could be opened, the factory had to reinstall its production line for forging magazines. To hold the floor plate closed and prevent the plate from opening accidentally, engineers placed a catch on the bottom of the magazine next to the trigger guard. They redesigned the military rear sight with a fixed aperture at 100 meters (109.36 yards) and an adjustable leaf sight graduated up to 500 meters (546.8 yards).

Production LK-62 series rifles retained the military rifle's "butterfly" safety, and the ejection port of those LK-62 rifles which were to be equipped with telescopic sights was lowered to prevent the ejected

Serbian and Yugoslav Mausers

cartridge case from hitting the scope. Sporting carbines were equipped with one of two trigger mechanisms, the double-set trigger or the military single-stage trigger.

While the LK series of sporting rifles were well designed and crafted, their limited choice of calibres available to buyers was a drawback. As they were based on the domestic Models of 1924 and 1948 with the 6 mm (0.236 inch) shorter bolt than the original Mauser Model 1898, they could only handle cartridge cases that did not exceed 80 mm in length (3.14 inches). This limited the choice of calibres to the 8 x 57 mm JS, the 7 x 57 mm, the 6.5 x 57 mm, the .308 Winchester (7.62 x 54 mm NATO), and the .243 Winchester.

A team of engineers worked to solve the ammunition problem. It took until 1967 for Djordje Matich, Milan Ilich, Milan Cirich, and Rodoljub Matkovich to develop a new system capable of accepting larger calibres, i.e., longer cartridges. The resulting LKK-67 carbine (LKK is the abbreviation in Serbian for "hunting carbine-short") incorporated the old receiver used in the Model 1948 rifle or the LK-62 or LK-62/64, but rechambered for either 8 x 57 mm JS, 7 x 57 mm, 6.5 x 57 mm, .308 Win, .243 Win, and the .22-250 cartridges plus other 6 mm ammunition.

The LKD-76 ("LKD" is the Serbian abbreviation for "hunting carbine-long") followed with a new receiver with a 78.5 mm long (3.09-inch-long) loading/ejection port and chambered for the more powerful 7 x 64 mm, .270 Winchester, and .30-06 Springfield ammunition. The same receiver was used in the LKM-67 (the LKM is the Serbian abbreviation for "hunting carbine Magnum") that was developed for such high-power calibres as the .300 Winchester Magnum, .264 Winchester Magnum, and 7 mm Remington Magnum.

The LK-67 series also introduced a new trigger assembly having a single adjustable trigger, a standard hunting-type safety located on the right side next to the bolt sleeve stop, and a magazine catch on the trigger guard.

Commercial sales of the LK-67 series failed to justify the labor involved in its manufacture and since its small-calibre choice limited its market, Zastava's management decided to stop trying to modify existing military patterns and develop an entirely new carbine. Work started with an analysis of the types of calibres available in the market, the size and weight of the best-selling sporting weapons, and the likely sales expected of different variants. The result was the LK-70 (Mark X) introduced in 1970, see Figure 11-1. Again, engineering selected the tried-and-true Mauser Model 1898 bolt assembly which was capable of handling large-calibre, high-power ammunition. A single receiver type but with three

Serbian and Yugoslav Mausers

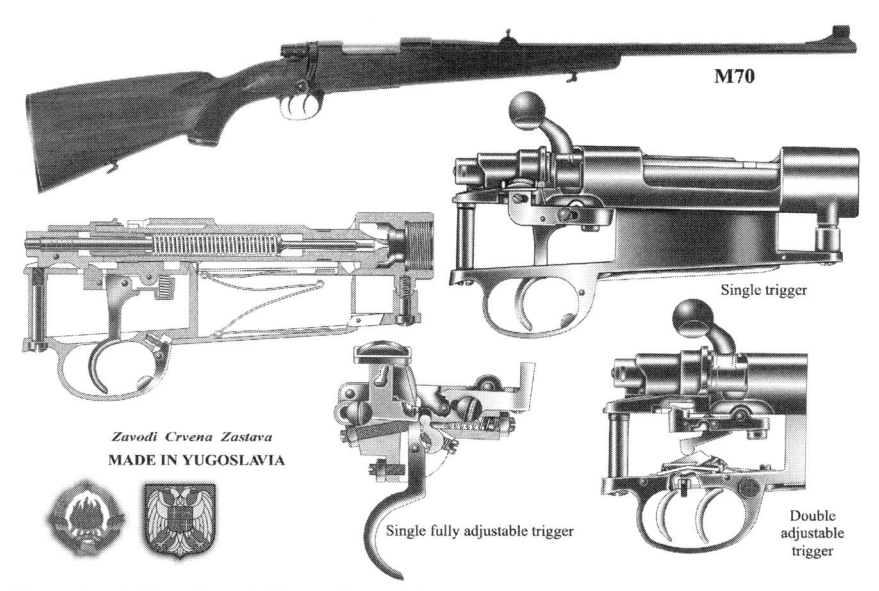

Figure 11-1. The Model 70 Hunting carbine.

loading/ejection port lengths was decided upon: 74 mm (2.913 inches) with an insertion to the rear for shorter ammunition, 84 mm (3.307 inches) for standard-length ammunition, and 94 mm (3.701 inches) with an extension on the front of the magazine to handle the longer Magnum cartridges. This increased the number of the calibres available to the potential purchaser to sixteen. The employees at Zastava literally felt they conquered the world when their Model 70 carbine began selling well in the United States, Great Britain, Germany, France, Austria, Belgium, Australia, Soviet Union, Denmark, Sweden, New Zealand, and other countries as well.

Equal in quality to the Model 70 carbine was Rodoljub Matkovich's Mini-Mauser Model 85 (a miniature action Mark X) made available in .223 Remington, .222 Remington, .222 Remington Magnum, .22-.250 Remington, .22 Hornet, and 7.62 x 39 mm calibres. He also designed a variant for left-handed shooters, the LK-85L chambered for the .223 Remington. In the late 1990s, Zastava produced the LK-22 Hornet, the LK-Long 65 (for .300 Winchester Magnum ammunition), the 99 Precision (for .22 Long Rifle and .22 Winchester Magnum ammunition), the MP22R, with the MP22SA semiautomatic variant (designed by Zoran Todorovich), the semiautomatic carbines LKP-66 and LKP-96A/B

Serbian and Yugoslav Mausers

and LKP-96C for .308 Winchester, 7.62 x 39 mm, and .223 Remington ammunition. Both carbines were adaptations of the military PAP Model 1959/66 and AP Model 1970 rifles. Finally, the Model 70 carbine was offered in 9.3 x 62 calibre with "express" sights.

APPENDIX A
SERBIAN AND YUGOSLAV MILITARY FACTORIES

The Military Technical Institute Kragujevac
—The Zastava Arms

The history of the Mauser system of arms is closely connected with the history of the Military Technical Institute (Vojno Tehnicki Zavod, VTZ) at Kragujevac, which has long been considered the cradle of the Serbian arms industry. Although this factory did not become involved with the manufacture of the Mauser system until 1907, its specialists and makers played a necessary and important part in choosing, inspecting and accepting arms and ammunition from foreign factories. VTZ began the licensed manufactured of Mauser system rifles in 1928, and to date, has produced or adapted close to two million rifles and carbines based on the various Mauser designs.

The Beginnings
The first factory building was the Handgun Repair Workshop, a "temporary" structure constructed about 1844 in Belgrade just beyond the range of Turkish artillery. Known as "Headquarters," it, and another facility, the "Workshop," were located on the slopes of the River Sava, an area known as Sava-mala IV, or the Sava district.

The specific site was located at the present streets, Prince Miloš, Resavska, Bircaninova and Nemanjina. The Headquarters was not particularly imposing as it was little more than a shed behind the "old" Military Academy, which is near the west wing of a later "new" Military Academy building.

The construction of the military facilities at Kragujevac started much earlier. Prince Miloš Obrenovich wrote on 12 April 1832, that he intended to "build large barracks and hospitals for soldiers at Kragujevac."

NOTE: Miloš Obrenovich (1780-1860) was Prince of Serbia from 1815 to 1839 and once again from 1858 to 1860.

The first building was 95 x 24.6 meters (312 x 81 feet), had two floors and was built of wattle and daub. It had sixty-six rooms and was completed in 1832. Shortly afterwards, he built a one-story, brick barracks building which measured 45.5 x 24.6 meters (149.3 x 81 feet) and had twenty-four rooms, one of which was a blacksmith shop called The Military Workshop. The barracks were built on a site 14,355 me-

Serbian and Yugoslav Mausers

ters square (47,096 square feet) located beside the Lepenica River, and enclosed by a high wooden wall. The first project at the new barracks was to recondition and assemble small arms, which was done by workers employed by the State Military Workshop.

The building of a new explosives depot and the manufacture of ammunition appeared to have been the priority. Until 1837, the powder plant was situated, according to a source, "in the town itself at the old building near the Prince's church (built in 1818) where an accident was easily to be expected."

A safer location was chosen in the area bordered by the right bank of the Lepenica river, and Gospodar's and Meta's Hills, next to the military cemetery. The new site saved the town from the "terrifying danger" posed by the ammunition plant. It was a two-storyed building and it was finished at the end of 1837. The new powder depot served to house powder and lead while weapons and accessories remained in the old building. During 1841, the government erected new blacksmith and locksmith shops, and between 1845 and 1847, it constructed a new two-storyed administration building. Also in 1847, a block of the military-craftsman buildings at Kragujevac, known as the *Oruzehranilnica* or Arsenal, were renovated and transformed into the Military Equipment Factory. A year later, Ilija Garasanin, the Minister of the Interior, advised the Council of State of the need to "found such kind of a plant capable of casting and manufacturing the military equipment." Prince Alexander Karageorgevich approved the Minister of the Interior's advice, leaving its final development to the Senate. Nevertheless, in spite of the fact that a center of domestic military industry was just about to be realized, in October 1848, the government decided that a new Gun Foundry would be built within Belgrade's Handgun Repair Arsenal. The Ministry of the Interior, however, informed the Council of State on 26 May 1849 that "in their opinion" a future Serbian armory should not be established as a commercial enterprise as proposed; instead they believed it would be less expensive if it was a military operation. The Senate and Prince Alexander Karageorgevich approved this proposal on 26 May 1849 and Belgrade's military-industrial complex was subordinated to the General Staff, in other words, to the army.

NOTE: Alexander Karageorgevich (1806-1885) was Prince of Serbia from 1842 to 1858.

Serbian and Yugoslav Mausers

The Kragujevac Era

The relocation of military factories from Belgrade to Kragujevac started a year later in 1850 and was accomplished in stages. First, on 13 August 1850 a steam machine originally intended for the Gun Foundry moved to Kragujevac. But the decision to relocate the entire Gun Foundry did not occur until 29 March 1851; consequently, the final transfer of all remaining machinery, and remaining semi-finished and manufactured products, was not completed until 31 July 1853.

In order to merge the military administration and gain control over the production capacities according to the decree by the Chief Artillery Office on 8 December 1855, both the powder production sections in Stragari and the workshops in Kragujevac were to be jointly managed.

Prince Miloš Obrenovich confirmed the decision and published *Organizing of the Main Military Arsenal Management* on 23 July 1860 as well as the decision on *Organization of the Armory Management at Kragujevac*. It was under these orders that "the military laboratory, caps workshop, arms and ammunition facilities and powder depots" were incorporated into the Main Military Arsenal. The new organization included the artillery and ammunition foundry and the carriage and boring workshop; after 22 September 1860, the Gunsmith Shop was included as well. Combined, they came under a single office titled, Armory Management. The Gunsmith Shop was officially recognized as a separate department that same day.

It was not until the publication of Regulation F. No. 1315 on 7 May 1862 that the Army established a centralized Artillery Management Office at Kragujevac. All military ordnance production came under this organization to include the powder mill in Stragari, the armory, the arsenal, the percussion-caps workshop, the powder depot, and the main laboratory. The Gunsmith Shop, however, was confirmed in the position of a separate, constitutional department of the Artillery Management Office on 5 June 1862.

During the period 1857-1878, The Gunsmith Shop had manufactured the Serbian Model 1867 Green-Lorenz rifle in 13.9 mm calibre rifle altered to a breechloading percussion system as well as the later Model 1870 14.9 mm calibre converted breechloading rifle, see Figures A-1 and A-2.

The function of the Gunsmith Shop as a plant for refurbishment and rework was confirmed by a resolution of the Artillery Committee in a session held on 11 February 1879 and subsequently by a meeting of the National Assembly on 31 May 1881. On 15 February 1883, the government issued a decree whereby the Artillery Management department was

Serbian and Yugoslav Mausers

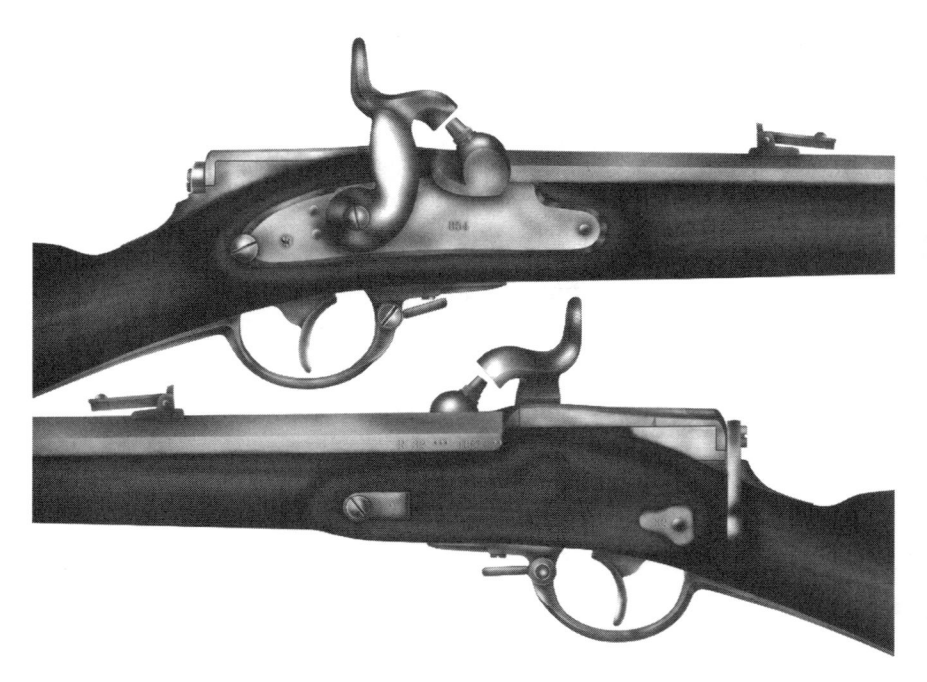

Figure A-1. The first Serbian breech-loading rifle, the Green alteration breech-loading percussion system, 13.9 mm Model 1867.

renamed the Management of Military Technical Institute of Kragujevac (VTZ). According to the table of organization dated 27 March, the Gunsmith Shop was ranked as a 3rd class workshop, a status that remained unaltered until 1914, see Figure A-3.

Figure A-2. The first Serbian breech-loading cartridge rifle, the Peabody conversion breech-loading cartridge system, 14.9 mm Model 1870.

Serbian and Yugoslav Mausers

In the course of the Balkan Wars as well as the World War I campaigns of 1914-1915, the Military Technical Institute operated at full capacity, which greatly contributed to the Army's battlefield successes. However, in the autumn of 1915, elements of the German 22nd Corps and the Austro-Hungarian 8th Corps of the III Koevess Army moved within striking distance of Kragujevac. Consequently, the Military Technical Institute initiated its evacuation plan on 15 October, and within less than a week it achieved a partial evacuation to Krusevac and Kraljevo. On 1 November, forces of the Central Powers occupied Kragujevac. Ultimately, the invading armies seized all of the evacuated machinery, equipment, and material and transferred thirty percent of it to Germany, thirty percent to Austro-Hungary, thirty percent to Bulgaria, and ten percent to Turkey.

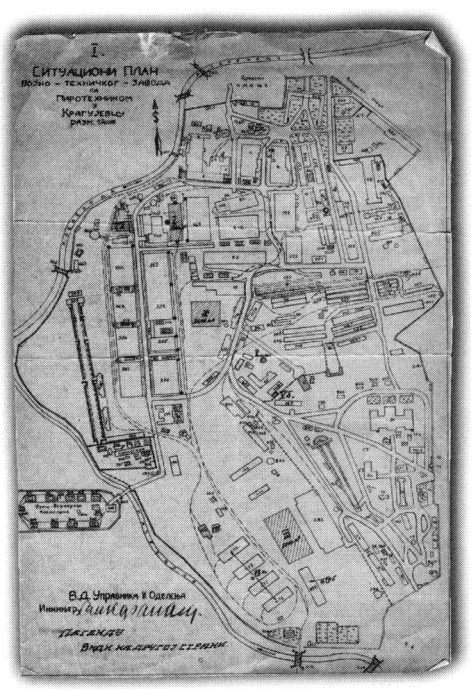

Figure A-3. The 1931 layout of VTZ Kragujevac.

ARTILLERY TECHNICAL INSTITUTE (ATZ)/ MILITARY TECHNICAL INSTITUTE (VTZ)

After the creation of a new state initially known as the Kingdom of the Serbs, Croats, and Slovenes but in 1929 renamed the Kingdom of Yugoslavia, the new government made a considerable effort to reestablish the Military Technical Institute (VTZ). On 30 November 1923 it was renamed the Artillery Technical Institute, or ATZ. Reconstruction began and, during the period 1928-1931, the ATZ gained a forging plant and some forging-die tool shops. In 1929, work finished on a fulminate plant as well, which gave the ATZ the ability to manufacture all types of initiating and detonating caps. Four years later, a new fuse shop was added as well. Meanwhile, on 4 March 1931, the expanded complex re-

Serbian and Yugoslav Mausers

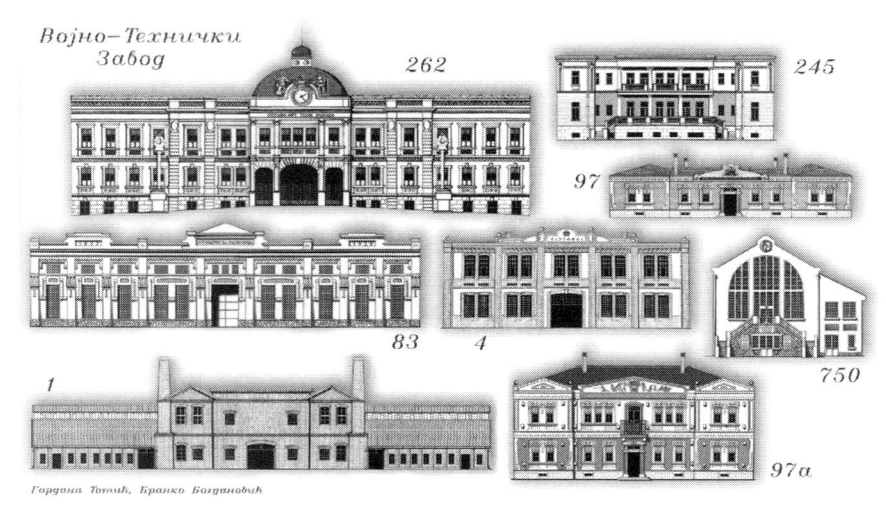

Figure A-4. The VTZ's buildings as of 1931. (262) Administrative building; (245) Manager's palace; (97) Old Military-craftsman school; (83) Artillery workshop; (4) Cartridge Case Workshop; (750) Electric-power station; (1) Gun foundry; (97a) Enlarged Military-craftsman school.

turned to its traditional name as the Military Technical Institute or VTZ. The final organization of the Military Technical Institute adopted on 1 January 1936 consisted of a central management office and 10 departments at Kragujevac with branches at Sarajevo, Skopje, Zagreb, Cacak, and Kamnik. See Figure A-4.

THE EFFECTS OF WORLD WAR II AND THE NAZI OCCUPATION
Shortly after the beginning of the Second World War but just prior to Germany's invasion of Yugoslavia, plans were made for the evacuation of the VTZ to Gornje Vogosce near Sarajevo if the need should arise. Unfortunately, nobody told the manager of the Military Technical Institute until 8 April 1941, two days after German operations had begun against Yugoslavia. The failure left little time to prepare for the move nor did the plan fully comprehend the enormity of the task. Only a partial evacuation of some machinery and the incapacitation of what remained had been achieved by 11 April 1941 when the forward elements of the German 11th Armored Division entered the VTZ complex. A short time later, the German administration, the Office of the General Economic Representative, or Amt des Generalbevollmächtigen fuer die Wirtschaft, headed by Franz Neuchausen and, more ominously,

Serbian and Yugoslav Mausers

the Military Economic Headquarters, Regensburg, for Special Tasks, assumed control.

Franz Neuchausen reported directly to Hermann Goering and he was responsible for the exploitation of Serbian economic capacities. Under a German four-year economic plan, with orders from the Wehrmacht's Military Economy and Armament Management Department or Wehrwirtschafts und Ruestungsamt OKW (WiRueAmt OKW), dated 28 April 1941, the Military Economic Headquarters became the Military Economic Headquarters for Serbia headed by a Colonel Braumueller. The next day, Colonel Braumueller began planning for adapting VTZ's factories into the structure of the German wartime economy. By the end of the month, the Germans classified the VTZ as being of special importance to the Wehrmacht. How this was to actually be accomplished was determined in a meeting held on 19 May at the main office of the Military Economic Headquarters. Attending the meeting were Franz Neuchausen, Major-General Rieder who actually ran the VTZ after 14 May 1941, and Waffen SS Obersturmbanfuhrer Alfred Baubin and a Major Wedel.

NOTE: After June 1941, the Military Economy and Armament Management Department was known as Military-Economic Headquarter Southeast.

In 1941, German industry was having difficulty producing reliable and safe mechanical time fuses ZtZS/30 (Zeit Zinder, 30 second). Based on a clock mechanism, the fuses were being mainly imported from the Swiss firms, Tavaro SA from Geneva and Oerlikon Buhrle and Company from Zuerich-Oerlikon. The German companies Gebr. Theil GmbH Ruhla, Gebr. Junghans, GmbH, Schramberg, and the clock and fuse industry in Schoenwald (a factory located in the Black Forest) also assembled the fuses from parts obtained from the Swiss clock manufactures Arnold Charpilloz, Fabrique Helios, Beveligrad, Vereinige Pignons-Fabriken AG, Grenchen, and Machines Dixi SA at Le Locle. It was a complicated system to operate, and having taken into possession the high-quality VTZ factory at Kragujevac, the Germans saw a possibility of improving fuse production as well as save 228,000,000 Swiss francs which could then be put aside to pay for finished fuses or the components for the ZtZS/30.

At a meeting held in Zemun it was suggested that the Military Technical Institute be incorporated into the Reichwerke Hermann Goering and initiate a month's production of 200,000 ZtZS/30 mechanical time fuses. The decision resulted in the assignment of Heeres Kraftpark 533 (or HKP 333) to the VTZ workshops. This commitment, however, did

Serbian and Yugoslav Mausers

not mean that there had been a change of mind by the Germans in regard to removing and exploiting of the captured raw materials and ready-made products found there. The Wehrmacht assigned Bergunskolonnen 333 from France to do the work. Bergunskolonnen 333 informed OKW on 14 and 16 June that among other things 250,000 remaining rifles of 7.92 mm Model 1924 and 5,000,000 rounds of live 7.92 mm Model 1924 ammunition had been taken from the VTZ to Germany.

THE RESISTANCE

Operations were soon affected by the growing Serbian resistance movement and by a labor boycott. Until the end of June 1941 the factory of Kragujevac received only one order for 35,000 ZtZS/30 fuses a month, leading the Germans to start thinking about what they needed to do to evacuate the whole factory to the Reich. On 17 August the department for the Military Economy and Armament Management at OKW sent the experts employed by Rochling Eisen-und Stahl Werke GmbH and a Mr. Simonis, the special advisor for military technology, to Kragujevac, to look into the feasibility of removing the tools and machines from the VTZ. Simonis was specifically responsible for determining which machines would be evacuated to Germany and which would be scrapped. He selected 2,500 machines for shipment to the Reich. Bergunskolonnen 333 managed the actual work and the unit evacuated the first portion of the equipment in November and December of 1941. The removal of the remaining machinery, equipment, raw materials, and finished material from damaged and destroyed buildings took until the end of 1943 to complete.

Most of the machines were initially moved to a staging area at the important railroad junction at Jasenice (Sammellager Assling). From there, they were transported by train to Kindeberg, 50 km to the north of Graz, and afterwards to the plants of F-bau company, Steyr-Daimler-Puch AG at Steyr, which was part of Reichwerke Hermann Goering. Only HKP-533 worked within the factory complex, and the old Cartridge Case Workshop was adapted to produce the Model 1924 7.92 x 57 mm ammunition.

According to the war diary kept by the Military Economic Headquarters for the Southwest, thirty-one cars of finished ammunition went to front-line units in July of 1942. In August, there was another shipment of thirty-five cars. During the second half of 1943, Military Field Workshop No. IV (Heeres Feldzeugpark-IV, HLP-IV) was assigned to what remained of the VTZ. It managed to get parts of the reconditioned infantry weapons

Serbian and Yugoslav Mausers

repair shops into operation. Nevertheless, when the Germans withdrew from Kragujevac they left the VTZ practically devastated. They took almost all machines, power installations, tools, raw material and semi-finished products, and they destroyed most of the buildings. Kragujevac was liberated on 21 October 1944 and the next day 300 former factory employees managed to find only 82 machines within the destroyed factory complex, and half of those were out of order.

After the Germans had left, it was quickly decided to reestablish production to support the National Liberation Army of Yugoslavia. To facilitate arms and ammunition production, the Supreme Headquarters of the National Liberation Army, which after 1 March 1945 became the General Staff of the Yugoslav Army, established the Military Industrial Department. One of the first initiatives of the Military Industrial Department was to rename the VTZ as the Weapons Factory of the National Liberation Army. The renamed VTZ was also organized along military lines and consisted of labor brigades, each representing a type of specialty. Each brigade had a military commander in charge. The brigades concentrated initially on repairing animal-drawn and motor vehicles and then they moved on to repairing artillery weapons and rebuilding the destroyed facilities and plants.

Post-World War II Period

On 26 April 1945, the Military Industrial Department became the Department for Military Industry and Supply; however, the president of the Antifascist Council of National Liberation of Yugoslavia decided as early as November 1944 to put all nationalized property under the control of the newly established Governmental Management for the National Goods, or DUND. The DUND quickly renamed the Weapons Factory as the Military Technical Shop.

The head of the General Staff, Lieutenant-General Arso Yovanovich, raised the question of managing the war industry. In his report to the Minister of Industry of the Democratic Federal Republic of Yugoslavia dated 20 May 1945, Arso Yovanovich wrote "that war production is a very unique industry working in support of the army, and in wartime it works for the front and, for those reasons, it should be put under a direct control of the Ministry of Defense." None other than the Prime Minister and Minister of National Defense, Marshal of Yugoslavia Josip Broz Tito, made the decision. He decreed in a top-secret order dated 20 June 1945 that eleven prewar factories and private industrial enterprises in the field of military industry including the important VTZ were to

Serbian and Yugoslav Mausers

be put under the control of the National Ministry of Defense. The conglomerate of industries that had been known as the Military Technical Shops was consequently renamed the Military Technical Institute on 28 July 1945. A subsequent order dated 31 December 1945 changed the title to the 21st October Military Technical Institute. By Order No. 58 of 4 September 1947 and by a decree from the government of the Federal National Republic of Yugoslavia, the organization received another title change to the Red Flag Enterprise (Preduzece Crvena Zastava) on 13 January 1948. The Ministry of Defense also introduced the Russian system of code designating material produced by this organization. This accounts for why a manufacturer's mark on the arms might appear as Institute 44 (Zavod 44), 28 July 1945 to 13 January 1945, or Enterprise 44 (Preduzece 44) for material produced after 13 January 1948 until 14 July 1962 when this system of code designating was discontinued. Post-Cold War socio-political changes caused the Weapons Factory to be renamed the Association of the Zastava Group, Zastava Namenski Proizvodi, DOO. The new organization got off to a good start as a consequence of overwhelming government funding and on 2 February 2002 the factory complex was renamed the Zastava Arms.

NOTE: Josip Broz Tito (1892-1980) was President of Yugoslavia from 1945 to 1980.

THE ARMS AND AMMUNITION FACTORY UZICE (FOMU) AND THE PRVI PARTIZAN UZICE (PPU)

As a consequence of a decision at the Paris Conference of 1919, the well-known armorer's centre Borovlje in Verlach belonged to the so-called "A Zone." In a plebiscite on 10 October 1920, fully 59.04 percent of voters voted for inclusion within what remained of Austria's territories and so the complex remained out of Yugoslav control. Pro-Yugoslav Slovenes were compelled to migrate from Austria to the new country, the Kingdom of Serbs, Croats and Slovenes, or SHS. Among others, the master gunmaker, Jakov Poshinger, came to Kranj in Slovenia. Along with other displaced gunmakers, he founded a guild named *Puskarna Kranj r.z.z.o.z* and became its first director. Also known as the Gunsmith Shop, it concentrated on the production of hunting arms. Jakov Poshinger, however, had bigger ambitions. He wanted to develop a military weapons factory in Serbia.

This would not be easy. In the immediate postwar period, Europe was flooded with surplus military weapons and even traders offering complete

Serbian and Yugoslav Mausers

weapons factories. Also, the defeated Central Powers were obliged to deliver, as a form of war reparation, a large part of their military industry to the Allies, including the Kingdom of the Serbs, Croats, and Slovenes. The country began receiving reparation material after the Spa Conference in July 1920 when Germany was forced into compliance with the Versailles Peace accords. Nevertheless, this material came in gradually for both Czechoslovakia and the Kingdom of the Serbs, Croats, and Slovenes. In the second half of 1920, the Gewehr und Munitions Fabrik Spandau-Berlin delivered part of its machinery to the Czechoslovakian factories Ceska zbrojovka v Strakonicich, Zbrojovka Praga v Vrsovicich, and Skodovy zavody, Plzen. The Statna zbrojovka v Brne received its machinery from the Vienna Arsenal and OEWG from Steyr, and a further 1,500 machines from the Mauser Werke AG. The Kingdom of the Serbs, Croats, and Slovenes received from this factory a total of 800 machines. Consequently, within a short period of time the military industry of Czechoslovakia and Yugoslavia thus gained the ability to produce the Mauser rifle Model 98 and 7.92 x 57 mm ammunition.

At the beginning of 1925, Statna zbrojovka from Brno introduced a new production line for the Model 1924 rifle production. Belgrade, however, had been offered disassembled Austrian OEWG, Wiener Arsenal, and Mauser factory machines from war reparation. The material was of good quality and still usable. Though manufactured under an American license from Pratt & Whitney and they were from the German plants Ludwig Loewe & Company, and the Deutsche Waffen-und Munitionsfabriken Berlin, A.-G (DWM), the Kingdom of the Serbs, Croats, and Slovenes had just bought up the license, installations, and machines for the production of the Model 1924 rifles from the Belgian company, Fabrique Nationale; consequently the Czech offer was rejected.

In contrast to the official attitude of government, Jakov Poshinger was aware that the worn-out production line from Brno was likely insufficient to provide the necessary arms and ammunition needed by the Yugoslav army. He personally did not have enough money to enter the negotiations with the Czechs for himself and he could not count on the state for financial support either, primarily because the government's treasury was exhausted by its investments into the Kragujevac factory. His solution was to sign a contract with Belgrade's Credit Union AD for a loan of 1,500,000 dinars.

NOTE: Belgrade's Credit Union AD (Beogradska zadruga AD) was founded in 1882 as the first Serbian insurance company with initial capi-

Serbian and Yugoslav Mausers

tal of 2.5 million golden dinars. The Brno factory was represented by the Czech Legion bank.

The new machines greatly improved Jakov Poshinger's factory's manufacturing capacity and, except for production of ammunition, the machinery was readily adaptable for producing replacement parts for Mauser rifles. This was an important development because the Army had adopted the Model 1924, which necessitated adaptation and unification of all patterns and systems having been in use until then to the Mauser system and to 7.92 mm calibre ammunition.

Jakov Poshinger needed to expand his factory and entered into an agreement with the Ministry of the Army and Navy to rent the barracks of the 4th Infantry Regiment Stefan Nemanja at Krcagovo in Uzice, Serbia. It took him until 25 October 1929 to complete a temporary installation of the machines, and the government's Court of Original Jurisdiction registered the factory as the Arms and Ammunition Factory Jakov Poshinger-Uzice, or FOMU, see Figure A-5.

Figure A-5. The barracks of the 4th Infantry Regiment Stefan Nemanja at Krcagovo, formerly the site of the Arms and Ammunition Factory (FOMU).

The project benefited by an important decision by the authorities of Krcagovo on 15 April 1935 to present Jakov Poshinger with a free building site in the town. He thus gained land for his factory and immediately set about erecting the necessary facilities. The Kingdom's Treasury

Serbian and Yugoslav Mausers

Ministry, in a confidential report (No. 215/VII) of 14 September, issued an official deed for the land to Poshinger on 10 October 1935. This was fortunate as the foundation stone for the first building, a 4,000-square-meter (43,030-square-foot) facility, had already been laid four months earlier on 20 June.

It was at this time that Jakov Poshinger decided to retire, and on 21 March 1936 he issued a statement before the district court at Uzice that he was selling ownership of the entire business to his son Jakov Poshinger Junior. The factory alone had an estimated value of 1,600,000 dinars ($145,454.54). The younger Poshinger was obliged to pay a debt to the Belgrade's Credit Union AD of 1,500,000 dinars ($136,363.63) plus interest. Some five months later, on 16 August the new owner went to district court and registered the business as the Arms and Ammunition Factory-Uzice-Jakov Poshinger Junior & Company. The younger Jakov Poshinger was a businessman, not a craftsman, and he instituted more aggressive business practices. On 27 November 1937 the *Official Gazette of the Kingdom of Yugoslavia* announced a transformation of the factory into a Stock Company with an investment value of 10,000,000 dinars ($181,818), of which 9,000,000 dinars ($818,181.81) which was the entire movable and unmovable properties of the factory, remained in Poshinger's hands while 1,000,000 dinars' ($90,909) worth of stock went into circulation. The business's new name became the Arms and Ammunition Factory of Uzice AD, see Figure A-6.

When Axis forces attacked in the Uzice area on 15 April 1941, the FOMU discontinued its production upon the entry of German units from the 8th Panzer Division into the factory complex in the early afternoon. The Germans did not immediately try to restart the production but for a time they oppressed even the owners of the factory based on nothing more than hatred of non-German peoples. Within a few months, the Germans started stockpiling the seized small arms and ammunition that had not already been taken to the collecting point at Pozega. The destruction by sabotage of other seized material at Smederevo, resulted in the seized material being placed in a meadow some distance to the east of the factory.

NOTE: It is quite possible that the Poshingers, like a considerable number of the gunmakers, had immigrated from Holland in the 16th century. Poshinger Senior is believed to have converted to the Greek Orthodox faith from Judaism. The Communists claimed that Jakov Poshinger Junior was a "Volksdeutcher" (an Aryan German national living outside the German national borders). As soon as the Occupation was established, he im-

Serbian and Yugoslav Mausers

Figure A-6. The Poshinger family in front of the FOMU factory.

mediately gave the factory to the Nazis and was afterwards incorporated into the German representative body for Serbia's economy. According to other records, Jakov Poshinger Junior had withdrawn from Uzice before the Partisans entered the town and returned with the Germans in late November or early December of 1941. He then sold part of the installation to Germany and part to the Independent State of Croatia (NDH). Having in mind Jakov Poshinger Senior's political and national adherence, the son's attitude is inexplicable.

    The approach of Partisan forces resulted in the withdrawal of German units from the area in the late evening of 20 September 1941. They had originally planned to leave the next day and their hurried departure made it impossible for German combat engineers to destroy the confiscated material as ordered. Concerned not to let it fall into Partisan hands, the senior German commander of the region entrusted the Luftwaffe with its destruction. They achieved some success in an attack on 22 September 1941 and destroyed most of the material. The Partisans entered Uzice on 24 September 1941 and a day later they possessed the practically undam-

Serbian and Yugoslav Mausers

aged factory. Within two days, they had gathered in the majority of the factory's work force and started production of *"Partizanka"* rifles and ammunition, as well as hand grenades and grenade launchers.

It was a short-lived effort. The Germans counterattacked and took possession of the factory again on 29 November 1941. A subsequent report by the Statistics, Industry, and Craftsmanship Department of the Ministry of National Economy (a part of Milan Nedich's Government), dated 5 December 1942, established that FOMU suffered "initially because of temporary work stoppage of 15 April 1941 and later because of destruction caused by the Partisans . . . During the course of Partisan disorders and German bombardment and after being plundered by the Partisans in October 1941," that part of the factory dedicated to arms production suffered 5,000,000 dinars' ($454,545) worth of damage to its buildings, 4,950,000 dinars' ($450,000) worth of damage to machinery and other installations, and the loss of 7,000,000 dinars' ($636,363) worth of raw materials.

Damage to the ammunition production plant was estimated at 3,000,000 dinars ($272,727), while damage to machinery and fixtures amounted to 9,400,000 ($854,545) dinars and the loss of raw materials amounted to 7,000,000 dinars ($636,363) for a total loss of 36,350,000 dinars ($3,304,000). In effect, FOMU was out of operation for the immediate future.

The Second Proletarian Division of the National Liberation Army of Yugoslavia, or NOVJ, liberated Uzice on 16 December 1944 and two days later most of the surviving workers gathered at the factory complex. Unfortunately, the extensive damage inflicted on the plant several years before made it impossible to return to immediate production. In early 1945, the factory was confiscated and put under control of the National Goods Management department, later known as the Ministry of Industry and Mining of the DFJ. In order to create a centralized government-controlled military industry, the DFJ decided on 20 June 1945 that the Ministry of National Defense would directly control the Arms and Ammunition factory- Uzice.

The FOMU had its name changed to the Military Technical Institute Uzice on 10 August 1945 and by the decree of the Treasury Ministry on 5 September 1947 it was renamed the "1st Partisan Tito's Uzice" or *Prvi Partisan B Titovo* Uzice (PPU). Four months later, on 23 January 1948, Tito declared the PPU to be a government-owned company of national importance dedicated exclusively to ammunition production.

Serbian and Yugoslav Mausers

THE MILITARY MAINTENANCE FACTORY, MILITARY WORKSHOPS

After 1945, the Mauser arms repair was done at a number of workshops of the Ordnance Corps, that is, the Technical Service of the Yugoslavian Army/Yugoslavian National Army (JA/JNA). Prior to the Second World War, the Kragujevac factory complex and its affiliated workshops in Sarajevo, Skopje, Zagreb, Cacak, and Kamnik were organized into ten departments, all of which were controlled by the Central Management of the Military Technical Department. The plants outside Kragujevac were not equipped to repair and modify sporting weapons or to manufacture the new Model 1924 7.92 x 57 mm Mauser rifles. As a consequence, when the war ended, the factory's military arms production departments and its maintenance sections had entirely different functions. In July 1945, the Artillery Section within the General Staff of the Yugoslavian Army was reorganized and retitled Artillery Headquarters. Within the new organization, the General Staff established the Ordnance Corps, which was responsible for fixed and mobile artillery maintenance workshops. The function of these workshops was to repair both artillery and small arms.

Among the many postwar changes imposed on the Army, the Ordnance Corps established a Principal Artillery Workshop in Zagreb, another Principal Artillery Workshop at Kragujevac, and Maintenance Workshop at Cacak. On 21 October 1950, the Principal Artillery Workshop at Kragujevac, which previously had been the pyrotechnics laboratory, was reorganized and renamed as Military Technical Workshop No. 515 or VTR 515. The new organization was in fact a part of the Red Flag Company, which had the code name Company 44, as the consequence of the Kragujevac factory having lost the rights of maintenance and ammunition production after a century of existence. VTR 515 was the de facto inheritor of the old Laboratory and IV Department of the VTZ.

NOTE: In July 1925, the General Staff established the Military Technical Institute, Cacak (VTZ. Cacak); In 1944, the Workshop for Supporting [military] Front and Inhabitants was founded; in 1945, the Maintenance Workshop was established; in 1947 and 1948, the Firm for Repairing Tool Machines and the Factory Boba Miletich, respectively, were established as well.

In 1953, the Ordnance Corps' Auto and Tank Technical services were combined and renamed the Technical Service of Yugoslavian National Army, or TS KoV JNA. The Technical Service decided in 1959 to discon-

Serbian and Yugoslav Mausers

tinue the Principal Artillery Workshop at Zagreb and expand VTR 515 to the level as the Maintenance Workshop at Kragujevac. The Maintenance Workshop at Cacak became budgetary Technical Overhauling Institution No. 1 (1.TRZ). Finally, in 1967, the TS KoV, the Technical Engineer Service and communication units, were merged and integrated into the joint Technical Service of the Armed Forces of Federal National Republic of Yugoslavia. The Technical Service possessed permanent, semi-permanent, and mobile workshops within military barracks, garrisons, and some territories. The combination of military workshops, maintenance plants, and military industry formed the country's military economic base. The permanent Class V technical workshops gave the military the capability for performing general maintenance of technical equipment and some spare-parts production. Model 1924/47 7.92 x 57 mm rifle assemblage and M98 rifle repair were accomplished at the Military Technical Institute 21st October of Kragujevac and in other permanent Class V technical workshops as well.

NOTE: 1.TRZ (Technical Overhauling Institution No. 1) Cacak was established to overhaul combat equipment, such as artillery and infantry weapons. It specialized during 1973 in converting the Mauser 7.92 x 57 mm into sporting rifles intended for commercial markets (income-based institution). For instance, the light carbine LK-3, derived from the Model 1948 7.92 x 57 mm rifle, sold in 1997 for $696.66 (1,045 dinars). With the 6 x 42 Zrak telescopic sight from Teslic, the price was $1,916.66 (2,875 dinars).

Codification of the workshops, maintenance plants, and factories of the military economic section was introduced on 14 May 1948. Unfortunately, the numerical list of the manufacturer's code-marks no longer exists, so only a few marks are known. See Table A-1 for a list of known markings.

Table A-1
Factory and Repair Depot Markings and Their Meanings
(These markings may be found on various Serbian, SHS, Yugoslavian Mausers)

RZK, 1, R-1 = Maintenance Institute No.1, Kragujevac (Remontni zavod Kragujevac);

1.TRZ = Technical Overhauling Institution No.1, Cacak (Tehnicki remontni zavod Cacak);

2.TRZ, R-2 = Technical Maintenance Institute No. 2 (unknown);

Serbian and Yugoslav Mausers

PREDUZECE 11, P11 = The 1st Partisan Uzice (Prvi partizan, Uzice);

PREDUZECE 12, P12 = Enterprise 12 (Igman, Konjic);

ZAVOD 44, Z44 (Institute 44) = 21st October Military Technical Institute, Kragujevac (Zavodi 21. oktobar, Kragujevac);

PREDUZECE 44, P44 (Enterprise 44) = The Red Flag Enterprise, Kragujevac (Preduzece Crvena Zastava, Kragujevac);

PREDUZECE 66, P66 (Enterprise 66) = Krusik, Valjevo;

VR. 69 = Military Repair Department, unknown;

RADIONICA 124 (Workshop 124) = unknown;

RADIONICA 145 (Workshop 145) = unknown;

TRZ. 136 = Navy Technical Maintenance Institute Tivat (?);

TR. 137 (unknown);

VTR 515, TRZ 515 = Military Technical Workshop No. 515, Kragujevac;

MBL = Milan Blagojevic, Lucani;

MOL = Marko Oreskovic, Licki Osik;

BK = Powder Plant Kamnik;

BK = Vojna Kontrola (Military Control) in use from 1947 to 1970;

PG = Pobeda, Gorazde;

RZN = Maintenance Department Neverke.

APPENDIX B
FINANCING SERBIAN MAUSER RIFLES

When Serbia gained complete independence from the Ottoman Empire in 1878, she found herself with few exploitable natural resources and a very small tax base. She also found herself in a very uneasy position vis-à-vis the Austro-Hungarian Empire and Bulgaria, both of whom would have liked nothing better than to incorporate Serbian territory. Organizing and equipping a first-class military force for self-defense was high on the priority list. But the government lacked the necessary resources to fund the purchase of the needed weapons, and Serbia was a long way from establishing the sophisticated manufacturing system described in Appendix A that was needed to produce their own military equipment. It was necessary for the Serbian government then to arrange foreign loans.

Figure B-1. The Mauser Arms Company factory at Oberndorf, Germany, in 1886.

FINANCING THE MODEL 1880 MAUSER-MILOVANOVICH
On 14 February 1881, Minister of War Milojko Lesjanin and Wilhelm Mauser agreed on a contract to deliver 100,000 Mauser Model 1880 rifles. The National Assembly quickly ratified the agreement on 26 February 1881. With the contract signed, Wilhelm Mauser, who had been living in Belgrade since 1879 working with the commission, felt that he could return to Germany to sort out problems at his firm Gebrueder Mauser & Company at Oberndorf am Neckar, see Figure B-1.

Serbian and Yugoslav Mausers

The crisis was largely financial but the Serbian contract valued at between 6,000,000 and 7,000,000 Reichs Marks (8,000,000 dinars or approximately $1.66 million) offered salvation. On his return to Oberndorf on 21 February 1881 he was greeted with loud cheers from the workers and citizens. The townspeople and workers, however, were unaware of the contract's details. What Wilhelm Mauser actually had were only the technical specifications; the awkward and difficult matter of financing on the part of the Serb government had not yet been resolved.

And resolution was not going to be easy as Serbia was seriously short of money. Nevertheless, on 25 May, when the government was in session, Serbian officials raised the issue of a 6,000,000 dinar ($1.2 million) loan. After some debate, the government approved the loan on 31 June and the Anglo-Austrian Bank, Vienna I, provided the funds. According to the agreement, Serbia had 15 years to repay the debt and the bank received a monopoly over all its salt exports during that period.

The loan agreement served as the basis for further negotiations between a Mr. Benzinger, a limited liability member of the Mauser factory, and the director of the Wuerttenbergischen Vereinsbank. In spite of Serbia's success with the Anglo-Austrian Bank, the difficulty remained Serbia's economic and political problems. A particular obstacle was forthcoming railway construction. The problem originated in November 1880 when Eugen Bontou, the Director of the French Union General company, showed considerable interest in railway business in Serbia. In January 1881, Eugen Bontou and the Serbian government reached agreement on a loan and signed a contract regarding construction and liabilities. The agreed-upon loan amounted to 100,000,000 French francs (100,000,000 dinars or $20 million) at an exchange rate of 71.4 percent including a 5 percent interest rate. This was a typical loan agreement in Europe at that time. The Serbian government subsequently issued 200,000 bonds each worth 500 francs (500 dinars or $100) with an interest rate of 25 francs (25 dinars or $5) to be paid off every six months. Eugen Bontou bought up the bonds at 317 dinars ($63.40) each with all debt to be paid off in 50 years. Serbia agreed to put the railway up as collateral as well as the incomes from both the railways and the customs as a guarantee. The Serbian government deposited the sum of the loan that was not immediately needed for railway construction with the Union General, that is, the Banque de France at an interest rate of 4.5 percent.

This arrangement suffered a heavy blow when on 19 January 1882 the Union General went bankrupt and Eugen Bontou was arrested. The Union General owed 42,900,000 dinars ($8.58 million) to Serbia subtracted by 8,400,000 dinars ($1.68 million) that Serbia had with the

Serbian and Yugoslav Mausers

Banque de France. Serbia needed the 34,500,000 dinars ($6.9 million) but the question was whether or not the bonds in possession of the now defunct Union General would be included as part of the Union General's debt and be equally shared among the creditors. Serbian Prince Milan Obrenovich pressed the Austro-Hungarian government to help Serbia out of the crisis without incurring further loss.

NOTE: Milan Obrenovich (1854-1901) was Prince of Serbia from 1868 to 1882, King from 1882 to 1889.

Thanks to the Viennese guarantees, the Comptoir National d'Escompte de Paris, or CNEP, Banque Nationale de Paris (the present-day BNP) appeared as a respectable negotiator. By a subsequent agreement with the Comptoir National d'Escompte, the huge loss threatening Serbia was avoided. Once past this crisis, an amount of 5,600,000 dinars ($1,120,000) was deposited in the account of the Wuerttenbergischen Vereinsbank. The total cost of the rifles ordered had been secured, but because of the financial delay, the Serbian government deferred arms delivery from 20 November 1883 to 13 January 1884.

Financing the Mauser Models of 1899, 1899/07, Model 1910 and Model 1908 Carbine

As the 19th century entered its final two decade, the Serbian Army had become concerned that improvements in small arms had outpaced their single-shot Model 1880 infantry rifle and Model 1884 carbine. Surrounded as they were on all sides by barely friendly or overtly hostile nations all armed with small-calibre, magazine-repeating rifles, the need to rearm with similar weapons had become critical.

But, as before, Serbia's financial situation forced the Army to defer its efforts to modernize and it was not until 28 April 1897 that the Cabinet was able to purchase 110,000 modern magazine rifles, plus one million rounds of ammunition. The Cabinet agreed to use credit from 1895 for the purchase with the remaining funds to be provided by a new loan. The Cabinet also approved the Minister of War's suggestion to spend 6,730,000 dinars ($1,346,000) on the rifles, 152,000,000 dinars ($30,400,000) on ammunition, and 1,000,000 dinars ($200,000) on ammunition pouches.

The Minister of the Treasury received authorization to sign a short-term loan of 4,000,000 dinars ($800,000) from the Serbian trade union of banks, dating from June 1895, with a six-month option. It had to use

Serbian and Yugoslav Mausers

local banks since Serbia was so encumbered with debt that it was unable to find a foreign bank to advance a loan, nor even a firm ready to deliver the arms on credit. Belgrade had expected the sympathetic cooperation of its usual supplier, Waffenfabrik Mauser, Oberndorf am Neckar, but Mauser was bound by an agreement with Turkey not to sell modern arms to its potential opponents . . .which included Serbia.

At the beginning of 1898, a new War Minister, Dragomir Vuchkovich, initiated a fresh examination of suitable rifles with which to modernize the Serbian Army. By May, it was again decided that the Spanish Model 1893 was the superior weapon. The process of selection was not entirely without political interference as the Mauser family was well connected with Serbian leaders. For example, Wilhelm Mauser's daughter Elisa had married General Kosta (Koka) Milovanovich in 1884. But, these connections notwithstanding, in June 1899, Mauser officials wrote to Michael Rashic expressing regret that the rifles for the Serbian Army had not yet been ordered from their factory.

The problem remained one of money. During the first half of 1898, the Serbian Minister of the Treasury, Vukashin Petrovich (1874-1924), tried unsuccessfully to get a loan from both the Berliner Handelsgesellschaft Bank and from Loewe & Company to buy arms. In July 1898, the government of Austro-Hungary promised Serbian King Alexander Obrenovich I that it would approve the loan. Armed with this promise, Vukashin Petrovich immediately traveled to both Berlin and Vienna to negotiate terms. But he quickly discovered that their conditions would prove to be unacceptable.

NOTE: Commanditgesellschaft auf Actien der Berliner Handelsgesellschaf'c at Behrenstraße 32/33, Charlottenstraße 33/33, Französische Straße 42, was founded in 1856; it merged with the Frankfurter Bank zur BHF, as late as 1970.

Austro-Hungarian bankers wanted, in return for a loan of 60,000,000 dinars ($12,000,000), all rights over Serbian railroads and forests. A group of banks from Berlin, in particular, the Bank fuer Handel und Industrie, managed by Hermann Bachstein, was ready to invest money in the construction of new secondary railways in return for the rights to use the main railroad. But, as it involved the exploitation of existing state railroads, approval by the National Assembly was required. A cabinet meeting held on 22 February 1898 chose to allow the Bachstein group only to construct and use 600 kilometers of new railways.

Serbian and Yugoslav Mausers

The terms were refused, thereby endangering the government's attempt to acquire the sum of 19,200,000 dinars ($3,840,00) necessary to start buying the desperately needed rifles. In fact, the contractor requested payment of 4,200,000 dinars ($840,000) by the end of 1899 and 15,000,000 dinars ($3,000,000) by August 1900. Once again, Serbian officials visited banks in Germany and Austro-Hungary.

In 1899, the Serbian Prime Minister received a message from Alexis Riese, the General Director of Deutsche Waffen-und Munitionsfabriken A.-G, in Berlin. Riese stated that, due to recommendations of Eugen von Minkus, the President of the Union Bank in Vienna, the factory was willing to enter into a contingency contract with Serbia to deliver 90,000 of the required 150,000 7 mm Mauser rifles, now designated the Model 1899, by Serbia.

NOTE: In 1884 Württenberische Vereinsbank sold the stock holdings of Gebrueder Mauser & Cie (Mauser Brothers and Company) to Ludwig Loewe & Co. On 6 February 1889 Ludwig Loewe bought Deutsche Metallpatronenfabrik as well as several other ammunition factories and on 10 October 1896 it took over the Mauser holdings from Loewe & Co. At the Karlsruhe Conference, on 4 November 1896 from these firms was formed the Deutsche Waffen-und Munitionsfabriken A.-G. On 4 April 1905, Deutsche Waffen-und Munitionsfabriken (DWM), Oesterreichische Waffenfabrik-Gesellschaft, Steyr (OEWG), Waffenfabrik Mauser, Oberndorf am Neckar, and the Belgian company, Fabrique Nationale d'Armes de Guerre, Herstal-lèz-Liège (FN), reached a cartel agreement with invested capital ratio of 32, 32.5, 20, and 15 percent.

Headed by Dr.Vladan Djordjevich, the Cabinet managed to raise a loan from the Union Bank, based on a nominal guarantee of 11,150,000 dinars ($2,230,000). Ultimately, the government received 10,032,971 dinars ($2,006,594.20), all derived from the gross income from the Serbian railways. The agreed rate of exchange was 87 percent of the total amount at a moderate interest of 5 percent to be paid off in 15 years.

While these negotiations continued, the Minister of War appointed an ordnance officer Nedeljko Vuckovich, as president of the commission formed to inspect and accept the completed rifles. A second ordnance officer was Damjan Vlajich. Production moved forward and by June 1899, the factory was ready with its first shipment of 100 rifles and 20,000 cartridges. The move was delayed two months due to the Austro-Hungarian government's hesitation in issuing a transit permit across Austria; however, by March 1900 Serbia had received 22,000 new rifles, which al-

Serbian and Yugoslav Mausers

lowed for the issue of 100 rifles per regiment. By June a further 39,000 rifles had arrived which allowed for the issue of 420 rifles per regiment. In 1900, the factory completed the full contract, see Figure B-2.

Figure B-2. Section No.VIII of the Steyr Arms Factory, Steyr Austria, at the end of the 19th century.

In 1909, the Serbian government again signed a loan agreement to procure additional small arms. The government approved another French loan for 150,000,000 dinars ($30 million) of which 95,000,000 dinars ($19 million) was set aside for the purchase of firearms. Serbia planned to purchase the remainder of the Model 1910 rifles and Model 1908 carbines from Oesterreichische Waffenfabrik-Gesellschaft, Steyr (OEWG) in Austria, but the government in Vienna suddenly imposed restrictions on meat imports from Serbia. In protest, the contract was granted instead to Deutsch Waffen-und Munitionsfabriken A.-G (DWM) for 32,000 Model 1910 Mauser rifles based on the German Model Gewehr Model 1898 rifle. Deliveries began that year with all 32,000 delivered by the summer of 1911.

APPENDIX C
RIFLE MARKINGS

Serbian arms did not begin carrying unique markings until the 1840s when Prince Miloš Obrenovic directed Dimitrije Radovich of the Viennese gun foundry, Geschuetzgiesserei, Wieden/Wien, Gusshausgasse, that the twelve Model 1759 3-pounder, 6-pounder and 12-pounder "Liechtenstein system" cannons that he had ordered be marked with the Serbian coat of arms. The Viennese banker, Teodor Tirke, charged Prince Miloš 780 florins for the work. It was not until 24 January 1856, however, that the Gun Foundry and Gunsmith Workshop at Kragujevac officially began stamping the firearms it produced for the Serbian army with the Serbian coat of arms.

RECEIVER MARKINGS

Each finished piece was stamped with a crest showing a shield with four steel strikers surmounted by a crown, the year of inspection, the ruler's cipher, and the controller's initials. The Gunsmith Workshop's first official controller, Mihailo Cvejich, traveled to Liege on 26 February 1857 to inspect and receive the shipment of 17.8 mm Minie-Francotte Petrovich Model 1856 rifles from the Auguste Francotte factory. Cvejich, as the inspector, stamped each rifle with his own initials "MC," the national coat of arms, and the year of manufacture (either 1856 or 1867).

Serbia continued to mark its military firearms in this manner until 1914, and then the only change made resulted in the manufacturer's name being translated into Serbian and stamped in Cyrillic on the receiver. From 1928 to 1941, domestically produced firearms also were marked with "phase control markings," that is, the numerals from 1 to 3 indicating which production line, the inspector's initials in Cyrillic, and the smokeless powder proof mark consisting of a crowned letter "T."

The method of stamping officially changed in 1945, though repaired arms continued to be marked without official approval. The Model 1924/47 rifles carried the Federal National Republic of Yugoslavia (FNRJ) coat of arms with five torches. The left side of the receiver was marked "Model 1924/47" and the name of the plant that assembled the rifle, for example, "ZAVOD 44" for Institute No. 44, "PREDUZECE 44" for Enterprise No. 44, "TRZ 5" for the Technical Maintenance Institute No. 5, and "RADIONICA 145" for Workshop No. 145.

Serbian and Yugoslav Mausers

Captured German rifles had manufacturer's markings removed but retained their original German inspection markings. The receiver ring was remarked with the FNRJ coat of arms and on the left side of the receiver was stamped in Cyrillic with the FNRJ factory acronym or name in the Latin alphabet, i.e., "PREDUZE 44" (for ENTERPRISE 44) that repaired the arm. Until production of the Model 1948 commenced, these marking were followed by the original pattern marks, either "M.98" or "Mod.98." From 1950 on, the factory stamped a preceding letter for those firearms converted according to a Yugoslavian model, for example the "M.98/48" and "Mod.98/48." These weapons also received new serial numbers as large numbers already had serial numbers prefixed with a Latin capital letter.

THE 1948 INSPECTION AND MARKING SYSTEM

With the advent of the Model 1948 7.92 x 57 mm rifle, the Yugoslavian army introduced a dual control system in the inspection process. Inspections occurred at all production phases within the factory. Military inspectors supervised the production process at its key points and inspected the finished weapon. A military commission was responsible for receiving the completed arms. Until the 1960s, the Land Forces, or KOV, controlled all military weapons production. In the early 1960s this responsibility transferred to the Army's Technical Department and in 1967 it moved again to the government's department for Military Industry Management. Technical maintenance plants, however, remained within the jurisdiction of the Technical Department of Armed Forces of the Yugoslavian National Army or OS JNA.

In these factories, controllers marked parts at the completion of certain points in the production process with the inspector's initials, the plant's mark which consisted of a letter inside a triangle and enclosing a letter and number, i.e., "K/13" inside, and a crossed-hammers proof mark after final assembly.

According to regulations, Model 1948 rifles bore the FNRJ coat of arms and marking "M.48" or "M48A" on the receiver ring. The receiver ring's left side carried the Cyrillic acronym "ФНРЈ" (FNRJ = Federal National Republic of Yugoslavia) together with the manufacturer's name "PREDUZECE 44" for Enterprise 44. The receiver and bolt assembly carried the serial numbers which were 3 mm (0.12 inch) high stamped with a hammer weighing 300 grams (0.66 pound). On the rifle's buttstock, the serial numbers were 10 mm (0.39 inch) high.

Serial numbers did not have a letter prefix. But serial numbers before circa 100,000 will often show a letter indicating the "batch" or "run,"

Serbian and Yugoslav Mausers

i.e., "T 88390" was rifle number 88390 in the Cyrillic "T" or 23rd batch. After circa 100,000, the batch letter designations were dropped.

According to technical specification, No. 21-1.23, only the barrel was proof-fired in a tunnel 5 to 8 meters (5.5 to 8.7 yards) long with a single Type II proof cartridge. If the barrel passed the proof firing and met all required tolerances, it was stamped with the letter "T." After the rifle was assembled, it was again fired with two Type I proof cartridges. After this trial, the bolt assembly, firing pin point, extractor and front notch were inspected. If they remained undamaged and met all specifications, they were stamped with the "T" mark as well.

REPAIRED ARMS MARKINGS

All repaired arms were inspected and stamped with a "T" surmounted by a five-pointed star if they met all specifications. A vertical line was stamped into the sight assembly following inspection of the sights. Finally, when the military commission accepted the finished weapon, it was stamped "BK" (VK) for Military Control—Military Commission.

THE BRUSSELS CONVENTION OF 1968 FOR NON-MILITARY FIREARMS

In its desire to meet international standards for civil arms, Yugoslavia signed the Brussels Convention in 1968. A year later, Brussels enacted rules governing the inspection and marking of arms and ammunition and, in 1970, its Institute for Inspection and Marking of Firearms and Ammunition began operations. The new regulations on inspection and marking changed barrel, receiver, and bolt markings on Yugoslavian weapons. They were now marked to show that the arm had undergone the specified proof firing for a finished arm, a final acceptance mark, and an inspector's mark. Thus, finished barrels were marked with the letter "T"; the mark of completed weapon was the acronym "BTB" and inspector's mark which now consisted of the person's initials surmounted by a five-pointed star.

If an existing barrel was reproofed, the old "BTB" marking was restamped but surmounted by a five-pointed star. Civil arms intended for legal export were stamped in the Latin alphabet. Arms intended for export to Third World countries had all markings indicating their country of origin removed.

See Tables C-1 through C-8 for all national, inspection and proof markings used on Serbian and Yugoslav rifles between 1880 and the end of Mauser production with the Model 1948.

Serbian and Yugoslav Mausers

Table C-1
National Coat of Arms
Serbia/Kingdom of the Serbs, Croats, and Slovenes/
Yugoslavia/Serbia and Montenegro
1899-2005

Coat of Arms	Marking	Country / Period
	КРАЉЕВИНА СРБИЈА	**KRALJEVINA SRBIJA** Kingdom of Serbia, (22. February 1882 – 1. December 1918.)
	КРАЉЕВИНА СХС	**KRALJEVINA SHS** Kingdom of Serbs, Croats and Slovenes, (1. December 1918 – 3. October 1929)
	КРАЉЕВИНА ЈУГОСЛАВИЈА	**KRALJEVINA JUGOSLAVIJA** Kingdom of Yugoslavia, (3. October 1929 – 17. April 1941)
	ДФЈ	**DFJ** Democratic Federal Yugoslavia, (10. August 1945 – 29. November 1945)
	ФНРЈ	**FNRJ** Federal National Republic of Yugoslavia, (29. November 1945 – 7. April 1963)
	СФРЈ	**SFRJ** Socialist Federal Republic of Yugoslavia, (7. April 1963 – 27. April 1992)
	СРЈ	**SRJ (F R Y)** Federal Republic of Yugoslavia, (27. April 1992 – 4. February 2003)
	С Ц Г	**SCG (S & M)** Serbia & Montenegro (4. February 2003 - ....)

NOTE: To determine a rifle's approximate period of manufacture, examine the Cyrillic markings on the receiver ring and compare them to the chart above. Then refer to the tables listing the serial number sequences for a particular model in the chapter describing that time period; i.e., a rifle marked "ФНРЈ" was manufactured between November 1945 and April 1963, indicating that it was probably one of the M48 series. Its serial number, 88390, indicates that it was manufactured in 1951 (Chapter 8, Table 8-1).

Serbian and Yugoslav Mausers

Table C-2
Kragujevac Factory Markings
1907-2003

ВОЈ.ТЕХ.ЗАВОД КРАГУЈЕВАЦ (emblem)		ВОЈ.ТЕХ.ЗАВОД КРАГУЈЕВАЦ	VOJ.TEH.ZAVOD. KRAGUJEVAC MILITARY TECHNICAL INSTITUTE, KRAGUJEVAC 15. February 1883 – 1. November 1915
BTЗ (emblem)		ВОЈНОТЕХ . ЗАВОД – КРАГУЈЕВАЦ	VTZ, VOJNO TEH. ZAVOD-KRAGUJEVAC, VOJ. TEH. ZAVOD-KRAGUJEVAC MILITARY TECHNICAL INSTITUTE, KRAGUJEVAC 1. December 1918 – 30. November 1923, and 4. March 1931 – 12. April 1941
		ВОЈ. ТЕХ . ЗАВОД – КРАГУЈЕВАЦ	
АТЗ (emblem) АРТ. ТЕХ ЗАВОД КРАГУЈЕВАЦ 1928 СЕДЛАРНИЦА		АРТ . ТЕХ . ЗАВОД – КРАГУЈЕВАЦ	ATZ, ART. TEH. ZAVOD-KRAGUJEVAC ARTILLERY TECHNICAL INSTITUTE, KRAGUJEVAC 30. November 1923 – 4. March 1931
44 З (emblem)		ЗАВОД 44	Z44, ZAVOD 44 INSTITUTE 44, KRAGUJEVAC 28. July 1945 – 4. September 1947
ПРЕДУЗЕЋЕ 44 (emblem)		ПРЕДУЗЕЋЕ 44	PREDUZECE 44, P44, CZ, ENTERPRIZE 44 (P 44), RED FLAG ENTERPRIZE (CZ), KRAGUJEVAC 4. September 1947 – 14. July 1962
44 (emblem)	7 (emblem)	PREDUZEĆE 44	
Z (emblem)	CZ (emblem)	Zavodi Crvena Zastava	ZCZ, Kragujevac RED FLAG INSTITUTE 14. July 1962 – 1. April 1991
(emblem)		Zastava Oruzje	Zastava Arms, Kragujevac, 22. October 2003, ……

Serbian and Yugoslav Mausers

Table C-3
Manufacturer's Markings
Receivers

Marking	Description
БР. МАУЗЕР и ДР. ОБЕРНДОРФ Н/н ВИРТЕНБЕРГ. ПЕШ. О.М. 1880.	BR. MAUZER I D^R. OBERNDORF $^N/_N$ VIRTENBERG: PES. O. M. 1880, Mauser Brothers & Co, Oberndorf, Wirthenberg, Infantry Breechloading Weapon M.1880
МАУЗЕРОВА ОР. ФАБ. ОБЕРНДОРФ н/н ВИРТЕМБЕРГ	MAUZEROVA OR. FAB. OBERNDORF $^N/_N$ VIRTENBERG, The Mauser Arms Factory, Oberndorf on /N, Wirtenberg
НЕМАЧКЕ ФАБРИКЕ ОРУЖЈА И МУНИЦИЈЕ БЕРЛИН	NEMACKE FABRIKE ORUZJA I MUNICIJE BERLIN, The German Arms and Ammunition Factory, Berlin
АУСТРИЈСКА ОРУЖНА ФАБРИКА ШТАЈЕР	AUSTRIJSKA ORUZNA FABRIKA STAJER, The Austrian Arms Factory, Steyr
ОРУЖНА ФАБРИКА МАУЗЕР А.Д. ОБЕРНДОРФ Н/н	ORUZNA FABRIKA MAUZER A. D. OBERNDORF $^N/_{N,}$ The Mauser Arms Factory SC Oberndorf on /N
(Tughra)	Far left: "Tughra" - Sultan Abdul-Hamid's personal signature; Left: "...Year. The Mauser Arms Factory SC, Oberndorf on /N" (top to bottom: Turkish Mauser M1903, M90 & M93).
(Mexican coat-of-arms) WAFFENFABRIK STEYR AUSTRIA	Far left: Mexican coat-of-arms (from 2. November 1821, to present days); Left: Arms Factory Steyr, Austria (Mexican Mauser M1912)

Serbian and Yugoslav Mausers

Table C-4
National Coat of Arms
and
Model Markings on Receivers

Coat of Arms Marking	Model Description
МОДЕЛ 1899	**MODEL 1899** Serbian infantry rifle 7 mm M99
МОДЕЛ 99/07	**MODEL 99 / 07** Serbian infantry rifle 7 mm M99/07
МОДЕЛ 1908	**MODEL 1908** Serbian carbine 7 mm M1908
МОДЕЛ 1910	**MODEL 1910** Serbian infantry rifle 7 mm M1910
МОДЕЛ 1899 С	**MODEL 1899 S** 7,9 mm carbine M99S - rebarreled Serbian infantry rifle 7 mm M99
МОДЕЛ 99/07 С	**MODEL 99 / 07 S** 7,9 mm carbine M99/07S - rebarreled Serbian infantry rifle 7 mm M1899/07
МОДЕЛ 1910 С	**MODEL 1910 S** 7,9 mm carbine M1910S – rebarreled Serbian infantry rifle 7 mm M1910

Serbian and Yugoslav Mausers

Table C-5
National Coat of Arms and Model Markings
Yugoslav and Domestic and Export Arms

Marking	Description
МОДЕЛ 1924	**MODEL 1924** 7,9 mm infantry rifle & cavalry carbine M1924
МОДЕЛ 1924 ЧК	**MODEL 1924 CK** 7,9 mm chetnik carbine – assault rifle M1924CK
МОДЕЛ 1924 Б	**MODEL 1924 B** 7,9 mm infantry rifle M1924B - rebarreled Mexican 7 mm Mauser M1912 & German 7,9 mm Mauser M1898
МОДЕЛ Л.J.	**MODEL L.J.** 5,6 mm sporting rifle Lazar Jovanovich
41.ĭ.	**Year 1941,** The First Partisan's Arms and Ammunition Factory, Uzice, September 24, 1941 - November 28, 1941. (**'Partizanka' rifle**)
У П О	**UPO,** **Uzicki Partizanski Odred,** Uzice Partisan Detachment. The First Partisan's Arms and Ammunition Factory, Uzice, September 24, 1941 - November 28, 1941. (**Partisan Rifle M1924B**)
	The Arabic alphabet: "**jim**", phonetic "**j**". Iraqi government property mark. The Iraqi National Congress opposition group has displayed this insignia in a context suggesting that it generically covers the Regular Army. This insignia and this usage is otherwise un-attested. (**Iraqi Mauser Rifle 7,9 mm M48**)
	Indonesian Police
	Syrian Coat of Arms 1947-1958

Serbian and Yugoslav Mausers

Table C-6
Various Markings Found on Serbian, Kingdom of Serbs, Croats and Slovenes, and Yugoslav Rifles

Marking	Description
MO IV	MO IV, Milan IV Obrenovich, 22. August 1854 – 11. February 1901. Prince of Serbia (1868 – 1882), King of Serbia (1882 – 1889)
(crown/A I monogram)	A I, Alexandar I Karageorgevich, 4. December 1888 – 9. October 1934. King of SHS/Yugoslavia (1921 – 1934)
(crown/P II monogram)	P II, Petar II Karageorgevich, 6. September 1923 – 3. November 1970. King of Yugoslavia (1934 – 1945)
(×ПК×)	PK, Accepted from Military Control (1924 – 1941)
Б	B, Inspector's Initials
Ш/T	Military Smokeless Powder Proof Mark, 1899 - 1915
(crown)/T	Military Smokeless Powder Proof Mark, 1924 - 1941
☆/T	Military Smokeless Powder Proof Mark, 1947 - 1970
(BK)	VK, VOJNA KONTROLA, Military Control (1947 – 1970)
△/п	P, Accepted from Military Control (1947 – 1970)
(BP)	VR, VOJNA RADIONICA, Military Workshop (1947 – 1970)
1.TRZ , (R gear)	1. TEHNICKI REMONTNI ZAVOD, "Technical overhauling institution" No.1, (Cacak, Serbia, 1944 – 2005)
ТРЗ5, TRZ.5	TEHNICKI REMONTNI ZAVOD 5, Technical Maintance and Repair Insitute No. 5 (Unknown, Yugoslavia, 1945 – 1991)
VR. 69	VOJNA RADIONICA 69, Military Workshop No. 69 (Unknown, Yugoslavia, 1945 – 1991)
△BP/24, ВОЈНА РАДИОНИЦА 124	VOJNA RADIONICA 124, Military Workshop No. 124 (Unknown, Yugoslavia, 1945 – 1991)
TR.137	TEHNICKA RADIONICA 137, Technical Workshop No. 137 (Unknown, Yugoslavia, 1945 – 1991)
△145, РАДИОНИЦА 145	RADIONICA 145, Workshop No. 145, (Unknown, Yugoslavia, 1945 – 1991)

Serbian and Yugoslav Mausers

Table C-7
Proof Marks
Yugoslavia
1970

#	Mark	Description
2	BTB	ORDINARY SMOKELESS POWDER PROOF
3	★ ★ BTB	REINFORCED SMOKELESS POWDER PROOF FOR FINISHED RIFLES
4	(crosshair symbol)	REPROOF MARKING FOR FINISHED BARRELS
6	★ BTB	FIRST SMOKELESS POWDER PROOF
7	T	STANDARD PROOF FOR FINISHED AND JOINED BARRELS
8	★ ★ T	DOUBLE PROOF FOR FINISHED AND JOINED BARRELS
9	★★★ T	TRIPLE PROOF FOR FINISHED AND JOINED BARRELS
10	PT	PROVISIONAL PROOF FOR UNFINISHED BARRELS
11	(pincers symbol)	ACCEPTING MARK
12	(crossed tools symbol)	ASSEMBLED RIFLES VIEW MARK
15	IO	MARK FOR RIFLES MADE FOR FOREIGN SALES - WITHOUT DOMESTIC PROOF
16	(anvil symbol)	SOLDER QUALITY MARK
18	Å K M̊ Š P̊ T̊	INSPECTION MARKS

BTB, Tormentacija Bezdimnim Barutom = Smokeless Powder Proof

T, Tormentacija = Powder Proof

PT, Provizorna Tormentacija = Provisional Proof

IO, Izvozno Oruzje = Export Arms

Serbian and Yugoslav Mausers

Table C-8
Principal Parts Inspectors
and
Their Markings
Models 1880 through Model 1924

Parts Inspectors Markings		
Initial	Name	Found on:
Б , Б	B, Brdarski Jovan (M1880)	M1880, M1924B
В , В	V, Vasiljevich (M1880) Vlajich Damjan (M1899, M1924)	M1880, M1899, M1924
Д	D, Danilo Barkovich	M1880
К , К	K, Kostich Kosta (M1880)	M1880, M1924
М	M, Markovich Milutin	M1880, M1884
Н	N, Najdanovich, Pavle	M1880
Р , Р	R, Rashich, Mihailo	M1880, M1899, M99/07, M1908, M1910
П	P, Petrovich, Milisav	M1884
С , С	S, Stefanovich	M1884, M1899, M99/07, M1908, M1910
Љ	LJ, Ljubisavljevich	M1880/07
Ј , Ј	J, Jovanovich	M1899, M99/07, M1908, M1910
Х , Х	H, Horstig, Pavle	M1899, M99/07, M1908, M1910
S	S, German Inspectors Mark	M1899, M1910

APPENDIX D
SERBIA'S WARS: A BRIEF HISTORY

WAR WITH TURKEY, 1912

Serbia made quick use of its newly acquired stocks of rifles. Beginning in February 1912, the government began preparations for a war against Turkey and provided such paramilitary troops as the *Komit* as well as Army and police units with the 7 mm Mauser rifle. Rifles also went to Albanian rebels in the provinces of Kosovo and Metohija allied with Serbia. The rifles were distributed by both border authorities and by the Serbian consul in Skopje.

The Albanian rebels regarded the Serbian Mausers highly, and they paid 16 to 17 Turkish lira (365 to 388 dinars or $73 to $78) for any rifle of this pattern marked with the Serbian coat of arms. They even went so far as to trade a good horse, a couple of oxen, and up to 30 goats or sheep for a smuggled Mauser rifle and as many as three boxes of ammunition. The amount of smuggled weapons was such that in July 1912 Turkish officers of the Kosovo *Vilayet* reported that they had reliable information to the effect that 24 packs of rifles, of which 660 were the 8 mm Mannlicher Model 1888/1890 pattern, had just arrived from Serbia.

Serbia provided the weapons to the rebels at no cost or, at least, for the same price as the government paid to Mauser. The rebellion's leaders bought needed weapons through the border towns of Vranje and Kursumlija as well as from Nish and Belgrade. For instance, on 29 July 1912, five Albanians waited for rifles sent by Belgrade. A Mr. Sejfudin from Seliste, Mr. Ahmet from Livac, and a Mita Filipovich agreed to buy the rifles through the Serbian consul in Prishtina. They crossed the Serbian border on the night of 30-31 July to buy 200 rifles for 100 Napoleons (100,000 dinars or $20,000). At the same time, another arms dealer, Casum Sefer, was busy selling Serbian Mausers to Albanian rebels at 6 lira (132 dinars or $26.40) each. In the regions of Turkish Macedonia, Kosovo, and Macedonia, other native Serbian rebel units equipped with the 7 mm Mauser Model 1899 and 8 mm Mannlicher Model 1888/90 were in operation.

Although the precise number of rifles sent to Albania and the number of weapons lost during the operations of both the *Komite* and the Serbian regular army during the campaigns of 1912 and 1913 is unknown, the number can be approximated.

Serbian and Yugoslav Mausers

Bulgarian War, 1913

Serbia's championship of Pan-Slavism in the Balkans engendered bitter rivalry with both Bulgaria and Austro-Hungary. Prince Milan, who had been proclaimed king in 1882, damaged Serbian prestige by entering an unnecessary and unsuccessful war with Bulgaria in 1885 over the question of Eastern Rumelia. Russia supported Bulgaria in this instance and Serbia found herself forced into an uneasy alliance with the Austro-Hungary Empire.

In 1912 the Balkan League had declared war on and defeated Turkey in the First Balkan War. But the League could not agree on division of the spoils and Bulgaria was extremely dissatisfied with the outcome. Serbia, sensing Bulgarian hostility, concluded an alliance with Greece. On 1 June 1913, King Ferdinand of Bulgaria ordered his army to attack Greek and Serbian positions in Macedonia. Fighting, principally in the Belles mountains and the Strymon Valley, was fierce. Bulgaria committed its 60,000-man army while Greece and Serbia fielded military forces of 100,000 and 80,000, respectively. Bulgaria was defeated after incurring some 18,000 casualties and a peace treaty (the Treaty of Bucharest) was signed on 8 August 1913. Greece and Serbia divided up the larger part of Macedonia between them. Serbia also gained the province of Kosovo.

The Second Bulgarian War, as it came to be known, would have far-reaching consequences for the future. Bulgaria was driven into an alliance with the Austro-Hungarian Empire and Serbia was driven ever closer to Russia and France.

World War I

At the beginning of the First World War in July 1914, the Kingdom of Serbia's Army reported a total of 131,391 7 mm calibre rifles in four models: Model 1899, Model 1899/07, Model 1910, and Model 1880/07, as well as 6,218 Model 1908 carbines in its possession. From this, it appears that approximately 67,609 rifles and 3,782 carbines went into the Albanian effort over a two-year period, refer to Table 1-1 at the beginning of the book.

Previous agreements with France and the fear of being absorbed into the vast Austro-Hungarian Empire were the deciding factors that forced Serbia to join the Allies in World War I.

The country was able to hold its own in the fighting that ensued until Bulgaria entered the war in alliance with the Central Powers. The scales were tipped against Serbia and the country was overrun. A government

Serbian and Yugoslav Mausers

in exile was established on the Mediterranean island of Corfu for the duration of the war.

Serbia's campaigns in 1914-1915 against the vastly superior Austro-Hungarian Empire had considerably reduced its stocks of weapons and ammunition. After the Central Powers prevailed in 1915, the Austro-Hungarian and Bulgarian armies seized a considerable number of Serbia's 7 mm Model 1899, Model 1899/07 and Model 1910 rifles. The arms were so well thought of that they were quickly adopted to their own specifications by the Austro-Hungarian Empire and designated the Model 1914 rifle.

Serbia appealed to her allies for help. At a conference held in Chantilly, France, on 12 March 1915, the Serbian Supreme Command representative, Colonel Petar Peshich, stated that Serbia now had only 20,000 Model 1899, Model 1899/07, and Model 1910 rifles remaining from the country's prewar stocks.

These weapons were repaired by the French firm, Manufacture d'Armes Saint-Étienne, and by the 1918 offensive were being used to arm recruits from the liberated regions. The prewar Mausers were held in such high esteem that following Germany's defeat they were used to arm the Serbian Guard Corps.

The Chief of the British Adriatic Mission, Lieutenant Colonel Willoughby Clive Garsia, had recommended to the Serbian government that it place orders for needed arms and spares with firms in the United States of America. Their immediate requirements amounted to 7,000,000 Model 99 clips, 160,000 Model 99 bayonets, 200,000 Model 99 rifle barrels, 60,000,000 brass plates for the production of cartridge cases, 50,000,000 brass plates for the production of bullet jackets, 200,000 Berdan primer caps, and 40,000,000 rounds of 7 x 57 mm ammunition.

But at an Allied conference in Rome held in early February 1915 it was decided that France assume the responsibility for supplying the Serbian army. The American order was consequently reduced to 160,000 bayonets, 1,000,000 Model 99 clips, and 70,000,000 cartridges. The bayonets were manufactured by the firm of Fayette R. Plumb, Co. of Philadelphia and St. Louis. The branch that manufactured the bayonets was located at Fayette R. Tucker Street and James Street, Philadelphia, Pennsylvania. The reduced Serbian order with the United States was handled by Mr. Gravath, the American representative at the Inter Allied Financial Council, who took responsibility for carrying out the contract.

Serbian and Yugoslav Mausers

WORLD WAR II

In March 1941, under intense pressure, Prince Paul, serving as regent for the young King Peter II, signed the Tripartite Pact with Nazi Germany and Italy. Hitler had two reasons for courting Serbia. First, the country provided an additional buffer to attack by the British from the south and secondly, it provided the Germany army with a corridor through which to attack the Soviet Union and gobble up the oil fields in southern Russia. But the pact was universally detested throughout most of the country; Prince Paul's government was deposed and the new government abrogated the pact.

The action so enraged Hitler that he delayed the opening offensive against the Soviet Union a crucial six weeks to attack Yugoslavia, and later Greece. German, Italian, Hungarian, and Bulgarian troops smashed into Yugoslavia from four directions. Belgrade was heavily bombed with great loss of civilian life and the country was overrun in six weeks.

Yugoslavia was dismembered. A German puppet state was created in Croatia. Dalmatia, Montenegro, and Slovenia were divided between Italy, Hungary, and Germany. Serbia came under direct German occupation and Macedonia became part of Bulgaria.

King Peter II established a government in exile in London and many units of the Yugoslav army fled into the mountains to conduct guerrilla warfare against Axis troops. Two major resistance groups developed. One was the nationalist *Chetnik*s under Dragoljub-Draza Mihajlovic and the second was the Communist Partisans led by Josip Broz Tito. At first, the two groups maintained a distant cooperation but by 1943, an open civil war had developed between the *Chetnik*s and Partisans. Tito was supported by the USSR and by Great Britain which provided extensive military aid and advisers. Without the support of the Allies, primarily Great Britain, the *Chetnik*s became more and more hesitant in their operations against Axis occupation troops. As a result, King Peter transferred the military command of the resistance to Tito.

In late October 1944, the Germans were driven from Yugoslavia by a combined offensive of the Partisans and the Soviet Red Army. Tito's national liberation council merged in November with the royal government. By March 1945, Tito had been elected premier and shortly afterward, the non-Communist members of the government resigned and were arrested. In November 1945, national elections gave the government to Tito by default as the opposition abstained from the elections. A constituent assembly was formed and proclaimed a federal people's republic.

APPENDIX E
THE GERMAN MODEL 1898 MAUSER

The German Model 1898 rifle reflected late-nineteenth-century technology that even before the First World War was obsolete in certain respects. For instance, the prototype Gew.88/97 had been chambered for 7.92 x 57 mm calibre Model 88 cartridge with an ogival-cylindrical bullet weighing 14.7 grams (226.8 grains) and loaded with 2.75 grams (42.4 grains) of Gew.Bl.P.88 (smokeless) powder. This produced a cartridge capable of a velocity (V25) of 620 m/s (2,034 ft/s) at 25 meters.

The barrel length was 740 mm (29.1 inches) and the complete rifle was 1250 mm (49.2 inches) long. Its rear sight (designed by Colonel Wilhelm Lange) had a fixed battle sight setting of 200 meters (218.7 yards) with the adjustable leaf graduated from 300 meters (328 yards) to 2000 meters (2,187.2 yards). Kaiser Wilhelm II's Cabinet was so impressed by the Gewehr 98, that it adopted the rifle as the service rifle for the entire German army. Field-testing began on 9 February 1899 with elements of 1. Garde-Regiment zu Fuss as well as the Berlin-based jaeger and rifle battalions and troops from the infantry school. The Gewehr 98 entered series production in 1900. The early Model 1898 was distinguished from previous Mauser models by its three locking lugs and by Colonel Lange's rear sight, a pistol grip, a butt-marking steel disc, and the small hook on the upper band which was used with the cleaning rod to "stack arms."

In order to equip the army as soon as possible, contracts were let simultaneously to the Prussian State arsenals in Danzig, Erfurt, and Spandau, to the Bavarian State arsenal in Amberg, and also to six private firms: Deutsche Waffen-und Munitionsfabriken A.-G (at Berlin-Charlotenburg and Berlin-Wittenau), HAENEL B C.G. Haenel Waffen & Fahrrad-fabrik AG (at Suhl, Thueringen), MAUSER B Waffenfabrik Mauser AG (at Oberndorf an Neckar, Wuerrtemberg), Schilling B V. C. Schilling & Co (at Suhl, Thueringen), Simson & Co, Waffenfabrik (at Suhl, Thueringen), and Kornbusch - Waffenwerke Oberspree Kornbrusch & Co (at Berlin B Hiederschoechweide).

Prewar modifications to the Gew.98 included, in 1904, the adoption of the "S" 7.92 x 57 mm ammunition with pointed bullet weighing 9.8 grams (151 grains). The bullet diameter was increased slightly. The powder charge was increased to 3.2 grams (49.4 grains of "Spandauer Pulver 682 b"). The bullet's velocity at 25 meters was 870 m/sec (2,854.3 ft/s) which generated 3679 Joules (2722.5 ft lb) of energy.

Serbian and Yugoslav Mausers

The round's larger diameter and its flatter trajectory necessitated the replacement of the chamber and the magazine and the rear sight. The new "S" rear sight had a fixed battle sight setting at 400 meters (437.4 yards) and was adjustable in elevation from 500 to 2000 meters (546.8 to 2,187.2 yards).

These changes were incorporated (order dated 1 October 1905) into the modernized Model 1898 (Gewehr 98/05). Converted Model 1898s were stamped with the Cyrillic letter "C" (Latin "S") on the receiver ring. On 19 November 1905, a further directive changed the round butt marking steel disk to the familiar "butt-eye," a metal plate with a hole in the center that served to provide an easy means of dismantling the firing pin mechanism. In 1915, other changes were made that included the addition of finger grooves and, after 1917, a change in stock wood from walnut to beech wood. The change was made as walnut was both more expensive and required a minimum of three years to seasoning before use.

Following World War II, the Model 1898 was adopted by a number of European and Asian nations, including Yugoslavia and Czechoslovakia. Political considerations in the wake of the Treaty of Versailles had led the two countries into close alliance. As detailed in Chapter 4, Czechoslavkia and Yugoslavia collaborated on the development of the Model 1924 7.92 x 57 mm Mauser, a shortened and improved version of the German Gew.98 rifle. The Model 1924, in turn, influenced the design of the German K98k that was adopted by Germany in June 1935 and used throughout World War II.

APPENDIX F
DISASSEMBLY/ASSEMBLY
SERBIAN AND YUGOSLAV MAUSER RIFLES

The Mauser Model 1898 rifle and its variations such as the Yugoslav Model 1924, 1927/27, Model 1952-Č, and Model 1948 were military firearms designed to withstand a great deal of abuse from hard usage and inclement weather over several decades. They were also designed to be easily disassembled for cleaning and general maintenance by individual soldiers. Follow the instructions given below and you will find your rifle very easy to disassemble.

Note: Always make certain that a firearm is unloaded before handling for any reason. Do not load a firearm until you are ready to shoot it. Always place the safety in the "ON" position when the firearm is not being fired, even if it is unloaded. Also, when disassembling the bolt, be aware that the firing pin spring is under significant tension. Always wear safety glasses and take care never to point the end of the bolt at your face or in the direction of another person.

Procedures for disassembly, reassembly, cleaning and maintenance of the various Mauser military rifles are nearly identical. Exceptions are noted.

To remove and disassemble the bolt:
1) Lift the bolt handle straight up and pull back enough to make certain that the chamber and magazine are empty. Figure F-1.
2) Push the bolt closed and turn down. Turn the safety lever at the rear of the bolt straight up.
3) Lift the bolt handle again and this time draw the bolt back.
4) With the thumb of the other hand, pull the bolt release on the left side of the receiver outward and draw the bolt out of the receiver. Figure F-2.
5) On the left side of the bolt, just below the root of the bolt handle is the bolt sleeve lock plunger. Being careful not to turn the safety lever from its vertical position, hold the bolt body in one hand and the bolt sleeve in the other, depress the bolt sleeve lock plunger, unscrew the bolt sleeve, and remove the firing pin, firing pin spring and bolt sleeve from the bolt body. Figures F-3 and F-4.

Serbian and Yugoslav Mausers

To dismount the bolt (Figure F-5):

6) If your rifle has the takedown bushing mounted in the side of the stock, insert the striker/firing pin in the bushing. Push down on the bolt sleeve until it clears the cocking piece lug. Figure F-6.

7) Holding the bolt sleeve down as far as it will go, turn the cocking piece one-quarter of a turn in either direction and remove it from the firing pin.

NOTE: The firing pin spring exerts a great deal of pressure on the internal bolt parts.

8) Carefully raise the bolt sleeve up the shaft of the firing pin to let the firing pin spring expand.

9) Turn the safety as far to the right as it will go and pull out of the bolt sleeve. Notice the cam on the bottom of the safety lever. It fits into and is held in a matching recess on the bolt sleeve.

10) Turn the extractor until the front end or claw is between the two locking lugs. Insert a screwdriver blade into the first opening behind the extractor claw. Figure F-7.

11) Carefully pry up the front of the extractor until the lug behind the claw is clear of its groove, then push forward and off.

12) DO NOT attempt to remove the extractor collar.

This completes the disassembly of the bolt.

To disassemble the stock:

13) Remove the cleaning rod.

Remove the magazine floor plate using a cartridge tip or other rounded object that will fit into the round hole in the floor plate just ahead of the trigger guard to depress the magazine floor plate catch. Slide the magazine floor plate back and lift off.

14) Remove the forward (H-shaped) barrel band. Depress the barrel band spring and slide the barrel band forward and off. Figure F-8.

15) Remove the rear barrel band in the same manner.

NOTE: The two barrel band springs on the Model 1924 and its variations are pinned into the stock. They can be removed by inserting a punch in the holes on the opposite of the forend and driving them out. Take care not to splinter the wood.

The barrel band spring on the Model 1898 and its variations is held in its channel by its own tension. When removing, take care that it is not lost.

Serbian and Yugoslav Mausers

16) Remove the trigger guard screws. Most Mauser trigger guard screws are held in position with lock screws, the head of which fits into a rounded cut in the head of the trigger guard screw to prevent them from backing out under recoil. Remove the lock screws first, then the trigger guard screws. Figure F-9.

17) Lift the barrel carefully from the stock. If it does not come easily, do not force it. Time, oil and wax have glued the metal to the wood. First tap gently all around the barrel channel with a rubber hammer. Then turn the rifle upside down and using a block of wood, gently tap on the edges of the magazine floor plate and the muzzle end of the barrel. If the wood and metal will not separate, daub lemon oil in the areas at the juncture of metal and wood and wait several hours. Have patience or you will splinter the stock. If the stock just fits tightly, you can turn the firearm right side up and gently tap the muzzle on a padded surface while holding the stock wrist. This will jar it loose. Figure F-10.

18) Remove the butt plate or butt cap by unscrewing the wood screws holding it in place.

19) There is no need to disassemble the trigger group. Instead, use a soft brush to clean away debris. Place a drop of oil of the trigger pin and wipe away excess.

NOTE: Further disassembly is not needed nor recommended for proper cleaning and maintenance.

Reassembly of the rifle stock is accomplished in reverse.

Reassembly of the bolt is easier if you know a few tricks.

20) To reinstall the extractor, stand the bolt body upright on a hard surface with the bolt handle pointing to your left. Place the extractor along the bolt body between the two lugs with the claw and its guiding lug overhanging the bolt face. Squeeze the extractor collar hard and push down on the extractor at the same time until it starts to seat. Holding the extractor in position with the fingers of your left hand, push up on the extractor claw with your right thumb until the guide lug clears the edge of the bolt face. Push back until the extractor guide lug can snap down into its groove.

21) Reinsert the safety lever into the bolt sleeve.

22) Slide the firing pin spring over the firing pin/striker.

23) Position the bolt sleeve over the striker. Place the cocking piece so that its projecting lug is oriented with the channel in the bolt sleeve and will slide down easily.

Serbian and Yugoslav Mausers

24) Either insert the firing pin/striker into the takedown bushing in the stock, or else mount vertically in a vise but do not clamp tightly. Holding the bolt sleeve in one hand and controlling the cocking piece with the other, push down on the bolt sleeve until the cocking piece can also be pushed far enough down to turn it one quarter turn in either direction to lock onto the striker.

25) Ease the bolt sleeve up until you can turn the safety lever to the vertical position.

26) Reinsert the bolt into the raceway, pull the bolt release lever out to the left, then depress the magazine follower to allow the bolt to move forward.

27) Turn the bolt handle down to lock it into the breech and either set the safety lever to the right, or turn it to the left to release it.

CAUTION: The bolt should be disassembled periodically and cleaned. When reassembling, wipe all surfaces with an oily rag. Do not overlubricate. If the rifle will be placed into storage for a lengthy period, coat the separate parts lightly with a good gun grease and store in a plastic bag. Attach the plastic bag to the rifle securely so they will not become separated. Before shooting the rifle again, disassemble the bolt and clean off the grease, then wipe all parts with an oily rag. Grease has a tendency to harden over the years and can possibly jam the firing pin forward, causing a slam fire when the bolt is closed on a live cartridge.

Serbian and Yugoslav Mausers

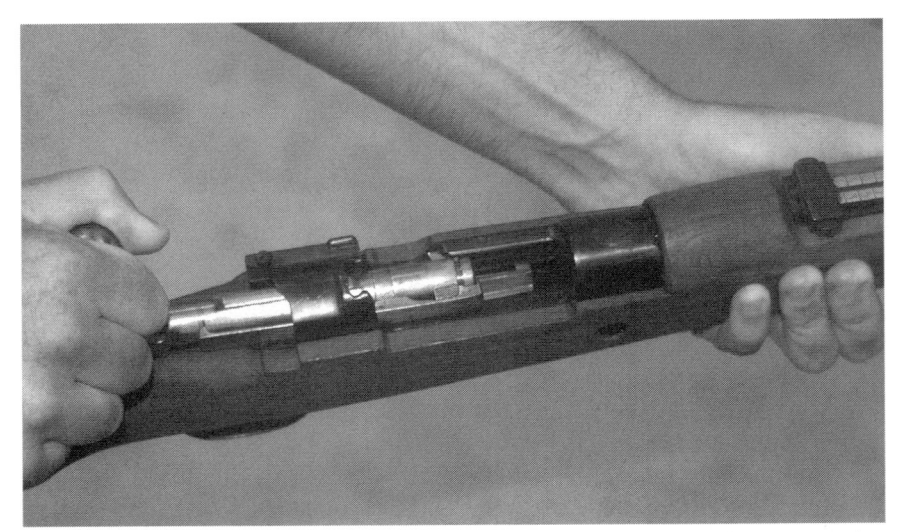

Figure F-1. Open the bolt to make certain the chamber and magazine are empty.

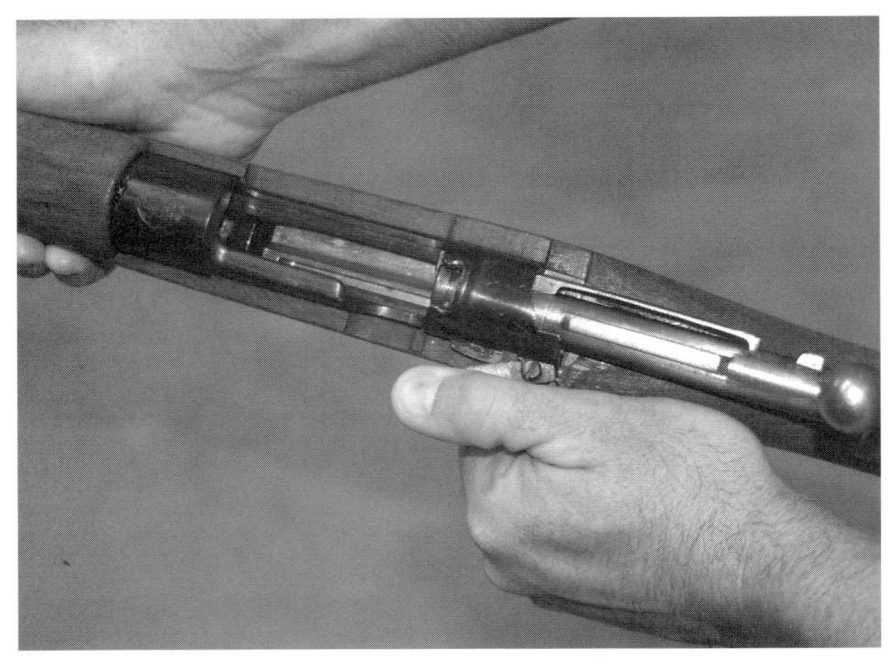

Figure F-2. Pull the bolt release lever out to release the bolt from the raceway.

Serbian and Yugoslav Mausers

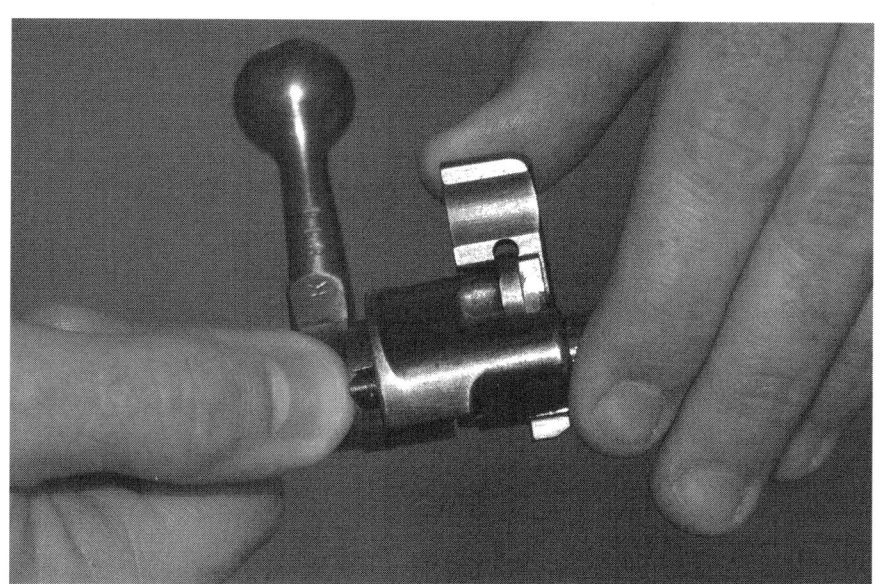

Figure F-3. Depress the bolt sleeve lock to unscrew the bolt sleeve from the striker/firing pin.

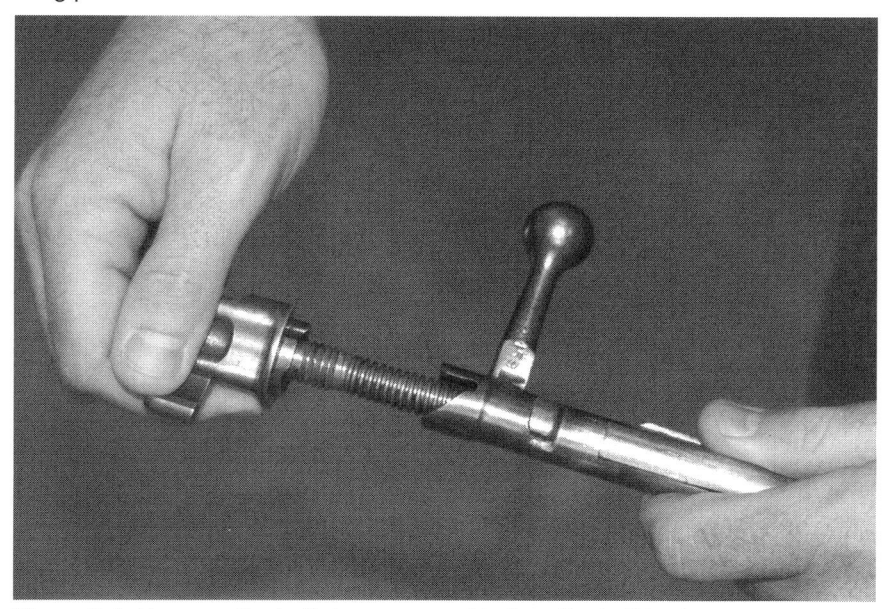

Figure F-4. Unscrew the bolt sleeve assembly from the bolt body.

Serbian and Yugoslav Mausers

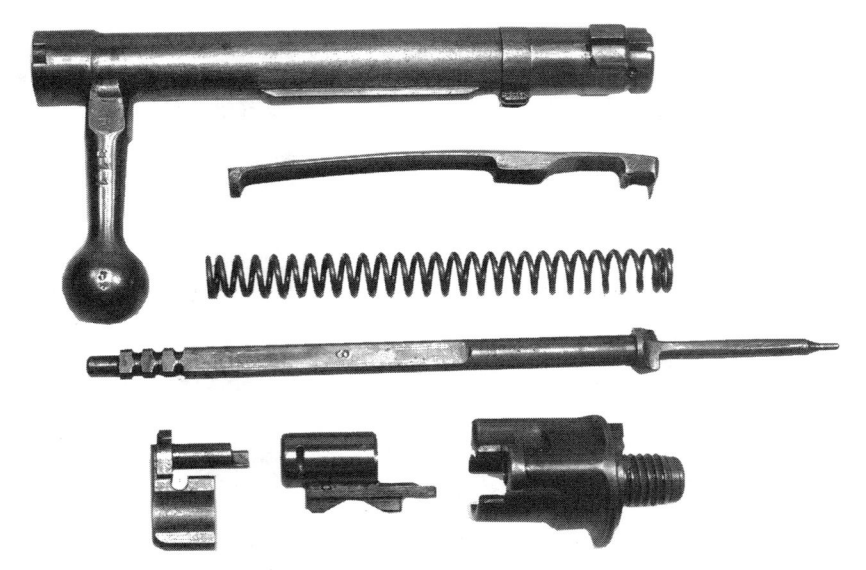

Figure F-5. Bolt Assembly for the Model 1924 7.92 x 57 Rifle. All Mauser bolts based on the Model 1898 Mauser design are similar.

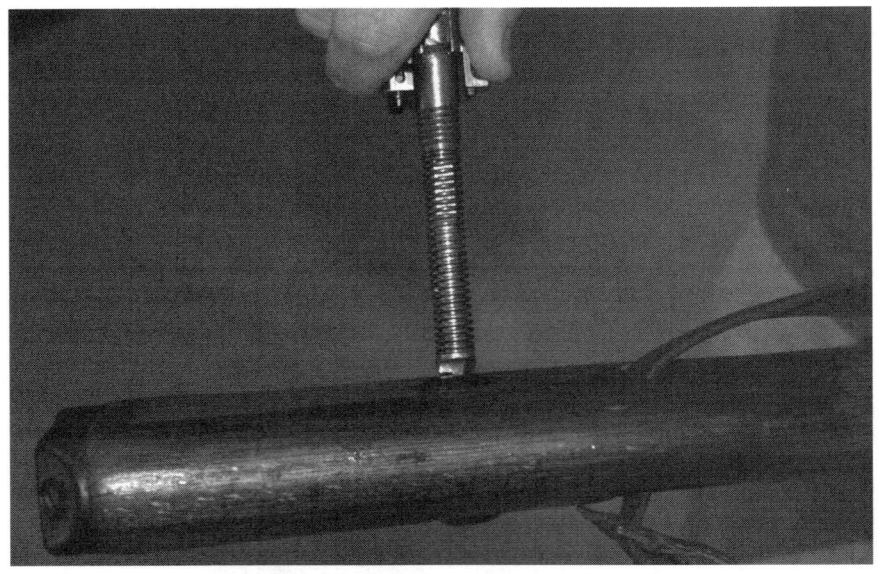

Figure F-6. If your rifle has a bushing in the stock for dismounting the bolt sleeve from the striker assembly, use that; otherwise, insert the firing pin end into a vise.

Serbian and Yugoslav Mausers

Figure F-7. To dismount the extractor, insert a screwdriver tip under extractor and pry up gently until the guide lug clears its groove. Push it forward and off.

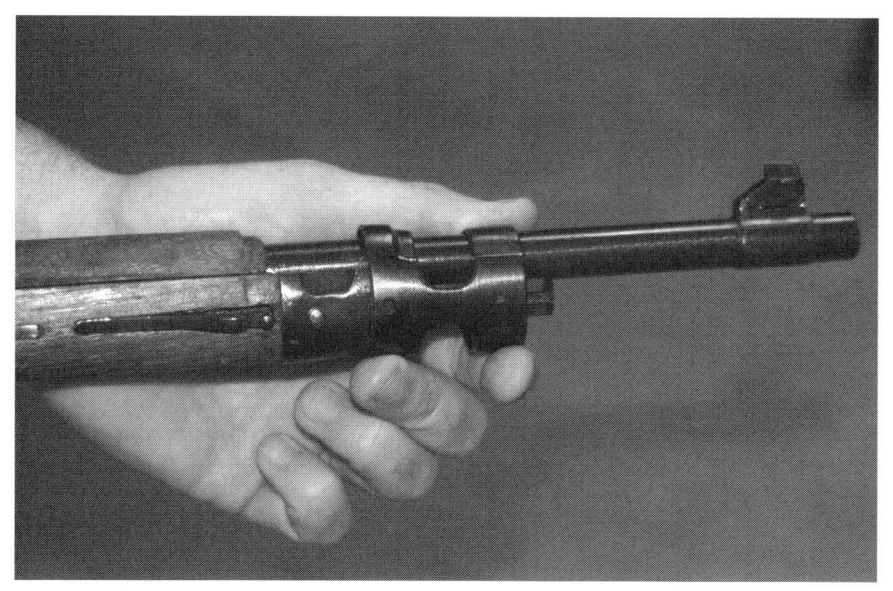

Figure F-8. Depress the front of the barrel band spring to disengage its stud and slide the barrel band forward and off the rifle.

Serbian and Yugoslav Mausers

Figure F-9. Trigger guard plate screws on Mauser rifles are usually held with lock screws to prevent them from backing out under recoil. Remove the small lock screw first, then the larger trigger guard plate screw.

Figure F-10. Remove the barreled action from the stock by lifting up and off. The handguard is held onto the barrel with clips. Remove by pulling straight up, carefully!

APPENDIX G
GLOSSARY

Chetnik
1. The term *"chetnik"* is used as a synonym for Royal Yugoslav special operations troops. It derives from the Turkish term, *çeté*, denoting guerrilla units.
2. Detachments of the Yugoslav Army of the Fatherland. Name given to several Serbian resistance groups in World War II organized to oppose occupying Nazis and their Croatian collaborators. Avoiding large-scale conflict with the invaders, the *Chetniks* mostly fought the Communist Partisans (q.v.) led by Josip Broz Tito. The most important *Chetnik* group was in Serbia, led by Draza Mihajlovic.

Cominform (Communist Information Bureau)
An international Communist organization (1947-56) including the Communist parties of Bulgaria, Czechoslovakia, France, Hungary, Italy, Poland, Romania, the Soviet Union, and Yugoslavia (expelled in 1948). Formed on Soviet initiative, it issued propaganda advocating international Communist solidarity as a tool of Soviet foreign policy.

Croatia
Formerly, part of the Kingdom of Serbs, Croats, and Slovenes and later Yugoslavia. Today, it is the Republic of Croatia with a population of 4.6 million in the northwest corner of the Balkan peninsula.

Cyrillic
Alphabet ascribed to missionary Cyril (ninth century), developed from Greek for church literature in Russian. Now the alphabet of the Soviet republics, Serbia, and Bulgaria, it is one of the three principal alphabets of the world.

Dinar
Serbian and Yugoslav national currency unit consisting of 100 paras. During the 125-plus years since 1880, the dinar has been subject to severe fluctuations in value due to wars and depressions and so it is impossible to establish the exact exchange rate for the dinar/U.S dollar. In this text, the following conversion rates were used:
 1880-1914: 1 USD = 5 dinars
 1918-1941: 1 USD = 55 dinars (approx.)

Serbian and Yugoslav Mausers

1945-1954: 1 USD = 11 dinars (approx.)
1955-1996: 1 USD = 1 to 1.5 dinars
1996-1997: 1 USD = 6.02 dinars
1997-1998: 1 USD = 5.99 dinars
1998-   : 1 USD = 10 dinars

Frequent devaluations have occurred since 1980. For instance, in 1980 the exchange rate was YD24.9 per US$1; in 1985 it was YD270.2 per US$1; and in 1988, YD2,522.6 per US$1. In 1990 new "heavy" dinar was established, worth 10,000 old dinars and the exchange rate that year was fixed at 7 dinars per West German deutsche mark. The revised rate in January 1991 was YD10.50 per US$1.

Dual Monarchy
Popular name for the Habsburg Empire after the 1867 Ausgleich (Compromise) that united Austria and Hungary under a common monarch; arranged to bolster the waning influence of Austria in Europe. Also known as the Austro-Hungarian Empire.

LCY (League of Communists of Yugoslavia)
Until 1990 the sole political party of Yugoslavia. Each republic and province had a separate organization, such as the League of Communists of Macedonia. Until 1952, it was known as the Communist Party of Yugoslavia (CPY). In 1990 the national organization split at the Fourteenth Party Congress; some republic Communist parties assumed different names, e.g., the Serbian party changed its name from League of Communists of Serbia to Socialist Party of Serbia. All the republic Communist parties remained intact (although reduced in membership) and ran candidates in the multiparty 1990 republic elections.

market socialism
The economic system introduced in 1963 in Yugoslavia based on worker-managed enterprises, using domestic and foreign market forces as a management guide.

nation and nationality
Juridically important distinctions that have played significant roles in Yugoslav political life, in spite of legislation that gave full equality to minorities in culture, public life, and language. The term *nation* was used in reference to ethnic groups whose traditional territorial homelands lay mostly within the modern boundaries of Yugoslavia, i.e., the Croats,

Macedonians, Montenegrins, Muslim Slavs, Serbs, and Slovenes. The term nationality, or national minority, designated groups in Yugoslavia whose homelands were outside Yugoslavia; the largest of these were the Hungarians and the Albanians.

pan-Slavism
A nineteenth-century intellectual movement that sought to unite the Slavic peoples of Europe based on their common ethnic background, culture, and political goals.

Partisans
Popular name for resistance forces led by Josip Broz Tito during World War II. In December 1941, the formal name, People's Liberation Army and Partisan Detachments, was adopted.

Serbia proper
The part of the Republic of Serbia not including the provinces of Vojvodina and Kosovo; the ethnic and political core of the Serbian state.

Slovenia
In the 19th century, Slovenia was part of the Austro-Hungarian Empire. It became part of the new nation, the Kingdom of the Serbs, Croats, and Slovenes (later Yugoslavia) in 1918. During World War II, Slovenia was occupied by Germany, Italy and Hungary. After World War II ended, it became a constituent republic of Yugoslavia. Slovenia declared its independence from Yugoslavia on 25 June 1991. After brief fighting, all federal Yugoslav forces withdrew the following month. Slovenia is now a parliamentary democracy.

Ustase (sing., *Ustasa*)
From the word *ustanak*, meaning uprising or rebellion. An extremist Croatian movement that began as an interwar terrorist organization, then adopted fascist guidelines and collaborated with German and Italian occupation forces in World War II. The movement's genocidal practices against Serbs, Jews, Muslims, and other minorities in Croatia and Bosnia and Hercegovina caused animosities that lasted long after the war.

APPENDIX H
BIBLIOGRAPHY

Babic, Branko. *Moskovke u naoruzanju crnogorske vojske*, GCM, knjiga III, Cetinje, 1970.

—. *Revolveri, sablje i nozevi u modernom naoruzanju crnogorske vojske*, GCM, knjiga V, Cetinje 1972.

—. *Ruski poklon berdanki Crnoj Gori*, GCM, knjiga I, Cetinje, 1968.

—. *Poceci savremenog naoruzanja crnogorske vojske, GCM, knjiga II*, Cetinje, 1969.

Bacalo, Branko. *Prljavo kraljevo zlato, feljton*, "Politika," Beograd, 1971.

Barker, A. J., John Walter. *Russian Infantry Weapons*, London, 1971.

Bjelajac, Mile. *Vojska Kraljevine SHS 1918 - 1921*, Beograd, 1988.

—. *Vojska Kraljevine Jugoslavije 1922-1935*, Beograd, 1994.

Bobrikov, G.I. *V Serbii - Iz vospominanii o voine 1877-1878*, s.Petersburg, 1891.

Bogdanovic, Branko. *Naoruzanje vojske u Prvom srpskom ustanku*, ZIMS br.21, Beograd, 1984

—. *Production of Firearms in FOMU from 1929 to 1941*, Uzicki zbornik broj 27, Uzice, 2001.

—. *TRENTE ANNEES DE PRODUCTION DE MUNITIONS DANS LA MANUFACTURE FOMU B PPU*, Uzicki zbornik broj 25-26, Uzice, 1996-1997.

—. *150th Anniversary of Arms Factory in Kragujevac (Kolevka srpske industrije B sto pedeset godina fabrike oruzja)*, Kragujevac, 2003.

—. *Arming of the Serbian and Montenegrin Armies from the 18th to the 20th Century*, Zbornik Istorijskog muzeja Srbije br.31, Beograd, 2003.

Serbian and Yugoslav Mausers

—. *Arms Factory in Kragujevac (Sto cetrdeset pet godina Zastava oruzja)*, Kragujevac, 1999/2000.

—. *Das Grosse Buch der klassischen Feuerwaffen*, Stuttgart, 1986 (I edition), Stuttgart, 1991, (II editon).

—. *Das Grosse Buch der klassischen Jagdwaffen*, Stuttgart, 1987 (I edition), Stuttgart, 1991 (II edition).

—. *Hladno oruzje Srbije, Cne Gore i Jugoslavije, 19-20 vek*, Beograd, 1997.

—. *Il Grande Secolo delle Armi da Fuoco*, Milano, 1987.

—. koautor, *Ministarstvo i ministri policije u Srbiji 1811-2001*, Beograd, 2002.

—. *L'âge d' or des ARMES DE CHASSE*, Paris, 1987 (I edition), Paris, 1991 (II editon).

—. *Le grand siècle des ARMES Á FEU*, Paris, 1986.

—. *RIFLES B Two Centuries of Yugoslavian Rifles*, Beograd, 1990.

—. *Savremeni pistolji i revolveri B rukovanje i odrzavanje*, Beograd, 1991.

—. *The Great Century of Guns*, Gallery Books, W.H. Smith Publishers Inc., New York 1986.

—. *The Great Century of Guns*, Tokio, 1989.

—. *Two centuries of the Police in Serbia (Dva veka policije u Srbiji)*, (I edition), Beograd, 2002.

—. *Two Centuries of the Police in Serbia (Dva veka policije u Srbiji)*, (II edition), Beograd, 2002.

—. *Les Pistolets meconnus de Lazar Jovanovic,* AMI No.78, Bruxelles, 1986.

—. *Pregled pesadijskog naoruzanja crnogorske vojske 1850-1900. godine*, GCM, knjiga IX, Cetinje, 1976.

Bolotin, D.N. *Sovetskoe strelkovoe oruzie*, Moskva, 1986.

Serbian and Yugoslav Mausers

Celap, Luka. *Austrijski vojni izvestaji o organizaciji srpske milicije 1821-1871. godine*, VVM br.10, Beograd, 1964.

Cetnik Sinisa. *Vojna snaga Turske, Srbije i Crne Gore*, Novi Sad, 1872.

Chamberlain, Peter and Terry Gander. *Axis Pistols, Rifles and Grenades*, London, 1976.

—. *Allied Pistols, Rifles and Grenades*, London, 1976.

Crna Gora i Srbija u 1862, Zapisi IV i V, 1929. 1935. godina.

Die Handfeuerwaffen des Osterreichischen Soldaten, Graz, 1985.

Dinic, Dragoljub. *Istorijski razvoj i postanak Topolivnice u Kragujevcu*, Vojnotehnicki glasnik br. 12, Beograd, 1953.

Djordjevic, dr. Vladan. *Crna Gora i Austrija 1814 - 1894*, Beograd 1924.

Djordjevic, Zivota. *Srpska narodna vojska 1861 - 1864*, Beograd, 1984.

Djurasinovic, Radomir. *Zastarela ratna tehnika*, Beograd, 1977.

Dolleczek, Anton. *Monographie der K.u.k. Osterr. ung. Blanken und Handfeuerwaffen*, Wien, 1896.

Dragicevic, R. *Poceci savremenog naoruzanja i organizacije crnogorske vojske*, Zapisi XXII, Cetinje, 1939.

Frilley, G., and Jovan Wlahovitj. *Le Montenegoro contemporain*, Paris, 1876.

Gavrilovic, A. Vojislav. *Istina o puskama,* Beograd, 1924.

Gluckman, Arcadi. *Identifying Old US Muskets, Rifles and Carbines*, Bonanza Books, New York, 1965.

Goetz, Hans Dieter. *Die deutschen Militaergewehre und Maschinenpistolen 1871-1945*, Stuttgart, 1981.

—. *Militaergewehre und Pistolen der deutschen Staaten 1800-1870*, Stuttgart, 1978.

Serbian and Yugoslav Mausers

Grujic, Sava. *Vojna organizacija Srbije s kritickim pregledom vojne organizacije starih i novih naroda*, Kragujevac, 1874.

Hicks, James E. *French Military Weapons 1717-1938*, N. Flayderman & Co., New Milford, USA, 1964.

Hitrova, N. I. *O ruskoj pomosci Cernogorii v period Vostocnogo krizisa 1875-1878*, Moskva, 1970.

Incognitus. *Srpska narodna misao i M. Pirocanac*, Beograd, 1895.

J.M.V, djeneral. *Prilog istoriji razvica srpske artiljerije*, Ratnik, sveska XI, godina.

Jovanovic, Slobodan. *Vlada Milana Obrenovica, knjiga II*, Beograd, 1927.

Juric-Keravica, Dobrila. *Oruzje iz partizanske radionice u Biokovu*, VVM br.6-7, Beograd, 1962.

Krenn, Peter. *Die Austruestung der steirischen Landwehr von 1808/09, "Die steirische Landwehr einst und jetzt,"* Graz, 1977.

Krpan, Ivan. *Istorija 45 pes. puka 1818-1922*, Maribor, 1926.

Lugs, Jaroslav. *Handfeuerwaffen I & II*, Berlin, 1973.

Lukic, Vasilije. *Krijumcarenje oruzja iz Austro-Ugarske preko Ulcinja za Albaniju*, GCM, knjiga V, Cetinje, 1972.

Lukovic, dr. Petko. *Nekoliko podataka o izlozenim puskama srpskih vojnika u Vojnom muzeju sa solunskog fronta*, VVM br.11-12, Beograd, 1966.

Martinov, B.P. *Rucno vatreno oruzje, sapirografisano*, Kragujevac, 1935/36.

Mathews, H.J. *Firearms Identification, Vol. 1*, University of Wisconsin Press, Madison, WI, USA, 1962.

Mavrodin, V.V., and Val. V. Mavrodin. *Iz istorii otecestvenogo oruzia*, Leningrad, 1984.

Milicevic, Milic. *Reforma vojske Srbije 1897 B 1900*.

Serbian and Yugoslav Mausers

Milovanovic, Kosta. *Artiljerija, Slike za artiljeriju*, Beograd, 1879.

Misic, Zivojin. *Moje uspomene*, Beograd, 1969.

Morin, Marco. *Le Armi Portatili dell'Impero austro-ungarico*, Firenze, 1981.

Naoruzanje kopnene vojsk, Beograd, 1963.

Novikov, V.I. *Opis municije za rucno vatreno oruzje, sapirografisano*, Kragujevac, 1936.

Operacije crnogorske vojske u Prvom svetskom ratu, Beograd, 1954.

Pastuhov, I. P. *Raskazi o strelkovom oruzii*, Moskva, 1983.

Pesic, Lj. *Snabdevanje srspke vojske zivotnim potrebama za vreme reorganizacije na Krfu*, Ratnik, VII, Beograd, 1939.

Pesic, Petar. *Pred pobedom*, Ratnik, VII, Beograd, 1925.

Petrovic, Radoslav. *Diplomatski spor o prenosu srpskog oruzja 1862. godine preko Rumunije*, Godisnjica Nikole Cupica, knjiga XLVIII, Beograd, 1939.

Pezo, Omer. *Opremanje naoruzanjem*, Razvoj oruzanih snaga SFRJ 1945-1985, Beograd, 1989.

—. *Vojna industrija Jugoslavije*, Beograd, 1983.

Pisarev, J.A. *Neki aspekti odnosa Rusije sa Crnom Gorom i Srbijom pocetkom Prvog svjetskog rata*, Istorijski zapisi br.2, Titograd, 1967.

Plenca, Dusan. *Oruzje iz partizanskih radionica u Vojnom muzeju JNA*, VVM br.1, Beograd, 1954.

Poleksic, Lj. *Kratak pregled istorijskog razvoja crnogorske vojske*, Ratnik br.X, Beograd, 1931.

Prvi balkanski rat 1912-1913. godine (operacije srpske vojske), Beograd, 1959.

Serbian and Yugoslav Mausers

Radenic, Andrija. *Austro-Ugarska i Srbija 1903-1918*, Dokumenti iz beckih arhiva, I, 1903, Beograd, 1973.

Radisavljevic, Slobodan. *Neki podaci o falsifikatima i drugim promenama na oruzju*, VVM br.11-12, Beograd, 1966.

Rakocevic, Novica. *Nabavka oruzja od strane Crne Gore u I svetskom ratu*, Istorijski zapisi br.I, Titograd, 1961.

Rankovic, J. Ziv. *BOMBE*, Stip, 1921.

Ristic, Jovan. *Spoljasnji odnosaji Srbije*, knjiga II, Beograd, 1887.

Rossia i nacionalno-osvoboditeljnaja borba na Balkanah 1875-1878, Moskva, 1978.

Sada, dr Miroslav. *Ceskoslovenske rucni palné zbrané a kulomety*, Prag, 1972.

Schmidt, Rudolf. *Die Handfeuerwaffen...*, *Atlas zu Schmidt*, Bern, 1878.

Schoen, Josip. *Montenegrinische Kriegsfuehrung und Taktik*, Wien, 1898.

Seitz, Heribert. *Kunigl. Armemuseum*, Stockholm, 1953.

Smith, W.H.B., and J.E. Smith. *Small Arms of the World*, 10th revised edition, The Stackpole Company, Harrisburg, PA, 1962.

Spasic, dr. Zivomir. *Kragujevacka fabrika oruzja 1853-1953*, Beograd, 1973.

Stojadinovic, dr. Milan. *Ni rat ni pakt*, Buenos Aires, 1963.

Stojicevic, M. Aleksandar. *Istorija nasih ratova za oslobodenje i ujedinjenje od 1912-1918 god*, Beograd, 1932.

Stranjakovic, dr. Dragoslav. *Vlada Ustavobranitelja 1842-1853. godine*, Beograd, 1935.

Terzic, Velimir. *Slom Kraljevine Jugoslavije 1941*, Beograd, 1982.

Turina, Milan. *Enciklopedija naoruzanja*, Beograd, 1958.

Serbian and Yugoslav Mausers

Vasic, Milos. *Specijalno i priguseno oruzje*, Beograd, 1988.

Venner, Dominique. *Les Armes Russes et Sovietiques*, Bordeaux, 1980.

Vinaver, Dr.Vuk. U*grozavanje Jugoslavije 1919-1932*, Vojno-istorijski glasnik, Beograd, 1968.

Vujovic, dr.Dimitrije-Dimo. *Crna Gora i Francuska 1860-1914*, Cetinje, 1966.

Vuksanovic-Anic, Draga, *Stvaranje moderne srpske vojske*, Beograd, 1993.

—. *Sa kapetanom d'-Omesonom 1877. 22 dana u dvokolicama kroz Srbiju*, Beograd, 1980.

Walter, John. *The German Rifle*, Arms and Armour Press, London, 1979.

Weeks, John. *Streljacko naoruzanje*, Zagreb, 1980.

Wolfgang, Zeel. *Mauser*, Zuerich, 1986.

Zdravkovic, Mileta. *Uzroci neuspeha Timocke divizije I poziva prilikom prelaska preko Save 6. septembra 1914.*

Zivanovic, Z. Milan, *Francuska i Srbija*, Beograd, 1970.

Zuk, A.B. *Revolveri i pistoleti*, Moskva, 1983.

—. *Vintrovki i avtomati*, Moskva, 1987.

Godin, Ratnik, sveska XI, godina L, Beograd, 1934.

Izvod iz vojnih zakona, pravila i propisa predvidjenih za izvodjenje nastave u zandarmeriji, knjiga IV, Novi Sad, 1939.

Ratnik, 1897 B 1941.

Sbornik zakona, uredaba i ostalih vojnih propisa trajne vaznosti u 1879. godini, Beograd, 1879.

Sbornik zakona, uredaba, naredjenja i objasnjenja vojnih izdanih u 1877. godini, Beograd, 1881.

Serbian and Yugoslav Mausers

Sluzbeni vojni list, 1879 B 1941.

Stenografske beleske Narodne skupstine 1897-1900; Uredbe i propisi za spremu i nosenje odela, Beograd, 1879.

Vojni zbornik zakona i uredbi za 1875. godinu, Beograd, 1875.

Zapisnici sednica Ministarskog saveta Srbije 1862-1898, priredio Skerovic dr, Nikola, Drzavna arhiva N.R.Srbije, Grada, knjiga II, Beograd, 1952.

Zbirka raznih zakona, uredaba, pravilnika, uputstava, naredjenja i raspisa vazecih za zandarmeriju, knjiga VII, Novi Sad, 1935.

Zbornik zakona, uredaba, naredjenja i objasnjenja vojnih 1858-1862. godine, Beograd.

Zbornik zakona i uredbi izdani u Knjazevstvu Srbije za 1863, 1864 i 1865. godinu, Beograd.

Zbornik zakona i uredbi izdatih u Kraljevini Srbiji 1897-1900, knj. 52 B 55; Vojna enciklopedija, br.1-10, Beograd, 1958-1967

Ilustrovana vojna enciklopedija, knjiga I (A- I), Beograd, 1939.

Merkblaetter ueber eigen und fremdlaendische Hand und Fraustfeuerwaffen, Wien, 1915.

Artiljerija, predavanja nizeg kursa Vojne Akademije u Beogradu za 27. klasu (1894-1897) i 28. klasu (1895-1898).

Borbena obuka u Sokolstvu, Savez Sokola Kraljevine Jugoslavije, Beograd, 1940.

Cenovnik tehnickih materijalnih sredstava KoV JNA, Beograd, 1970.

Pesadijska nastavna sredstva, Dinic, Obrad, Radenkovic, Milorad, Beograd, 1971.

Naoruzanje pesadije, Dinic, Obrad, Radenovic, Milorad,

Serbian and Yugoslav Mausers

Imenici pesadijskih sredstava i pesadijskih kompleta, sastavnih delova pesadijskih kompleta, nastavnih pesadijskih sredstava i sastavnih delova nastavnih pesadijskih kompleta, Beograd, 1962.

Imenici pesadijskih sredstava i pesadijskih kompleta, sastavnih delova pesadijskih kompleta, nastavnih pesadijskih sredstava i sastavnih delova nastavnih pesadijskih kompleta, Beograd, 1984.

Izvestaj o komisijskom radu za izbor novog puscanog modela za naoruzanje kralj. Srpske pesadije, sa 59 tablica i slika u osobenom atlasu, Izdanje Ministarstva vojnog, Beograd, 1886.

KARABIN Mauzer M.24a, kalibra 7,9 mm, Izvod iz vojnih zakona, pravila i propisa predvidjenih za izvodjenje nastave u zandarmeriji, knjiga IV, drugo izdanje, Novi Sad, 1939.

Nasa nova puska M. 24. (sa slikama u tekstu), priredio Vudjan, T. Branko, Sarajevo, 1926.

OPIS BRZOMETNE PUSKE Model 1898A 1907.-15. GOD, MOD. 1895. GOD. i MOD. 1890. GOD, Beograd, 1928.

Opis cuvanja i ciscenja srpske brzometne puske kalibra 7 mm, modela 99, Beograd, 1904.

Opis karabina 7 mm M.8.S, Beograd, 1932.

Poluautomatska puska M59-66 M59-66A1 i M59, knjiga II, (opis, radionicko odrzavanje i remont), Beograd, 1968.

Potrebna znanja za sokolske streljacke otseke, Savez Sokola Kraljevine Jugoslavije, Beograd, 1939.

Potsetnik za rezervne konjicke oficire, strucni deo, knjiga I, Beograd, 1937.

Potsetnik za rezervne pesadijske oficire, Beograd, 1938.

Pravilo puske mauzer M.98 i M.24, Beograd, 1927.

Puska 7,9 mm M48 sa tromblonom M.60, Beograd, 1968.

Serbian and Yugoslav Mausers

Srpska brzometna puska kalibra 7 mm, modela 99, Beograd, 1900.

Arhiv muzeja Zastava, Kragujevac

Arhiv Srbije, Beograd

Arhiv Vojske Jugoslavije, Beograd

Vojni muzej, Beograd

Serbian and Yugoslav Mausers

About the Author, Artist, Translator, and Editor

Branko Bogdanovic is a researcher and writer for the Zastava Arms Factory, Kragujevac, and a member of the Advisory Board of the Military Museum in Belgrade, Serbia. He is a professional associate of the Ministry of the Interior. He is also officer-in-charge, Special Antiterrorist Unit History and Tradition Division.

Branko Bogdanovic was born in Belgrade in 1950. He studied at the University of Belgrade and has published numerous articles and papers in the field of arms and military history in European professional journals. His published books include: *History of Zastava Arms* (Kragujevac, Serbia, 2003), *Das Grosse Buch der klassischen Feuerwaffen* (Stuttgart, Germany, 1987, 1991). *Il Grande Secolo delle Armi da Fuoco* (Milan, Italy, 1987), *L'âge d' or des ARMES DE CHASSE* (Paris, 1987, 1991), *Le grand siècle des ARMES Á FEU* (Paris, 1986), *The Great Century of Guns* (New York, USA, 1986; Tokyo, Japan, 1989), *History of Serbian Police Forces* (Belgrade, Serbia, 2002), *Two Centuries of Yugoslavian Rifles* (Belgrade, 1990), *Swords of Serbia and Montenegro* (Belgrade, Serbia, 1997).

The author, Branko Bogdanovic (kneeling, facing front), during recent antiterrorist exercises.

Gordana Totic is an architect and a professional illustrator in the field of arms history. Mrs. Totic has illustrated other books for Mr. Bogdanovic including *History of Zastava Arms* (Kragujevac, Serbia, 2003), *Das Grosse Buch der klassischen Feuerwaffen* (Stuttgart, Germany, 1987, 1991). *Il Grande Secolo delle Armi da Fuoco* (Milan, Italy, 1987), *L'âge d' or des ARMES DE CHASSE* (Paris, 1987, 1991), *Le grand siècle des ARMES Á FEU* (Paris, 1986), *The Great Century of Guns* (New York, USA, 1986; Tokyo, Japan, 1989), *History of Serbian Police Forces* (Belgrade, Serbia, 2002), *Two Centuries of Yugoslavian Rifles*, (Belgrade, 1990), *Swords of Serbia and Montenegro* (Belgrade, Serbia, 1997).

Serbian and Yugoslav Mausers

Branka Milosavljevic is currently a senior curator at the Military Museum in Belgrade. She received her degree from the Faculty of Philosophy at the History Department of the University of Belgrade. Since 1981, she has served as curator in charge of the arms collections from the 16th century to 1918. Mrs. Milosavljevic has organized numerous exhibitions, cataloged the museum arms collections as well as written several articles, and participated in two international professional symposiums. Her best-known catalogs are: *Swords in the Military Museum, XIV-XX* (Belgrade, Serbia, 1993) and *European Small Arms, XVI-XIX* (Belgrade, 1991).

Dr. Charles H. Cureton is Chief, Museums and Historical Property at the United States Army Training and Doctrine Command. His career in museums began in 1983 following completion of graduate work leading to a doctorate in history and he has been in his current position since 1988. He has also served in the U.S. Marine Corps as a reserve officer with the Marine Corps Historical Center and retired as a lieutenant colonel following such assignments as historian with the 1st Marine Division during the Persian Gulf War (1990-1991), with Joint Forces in Somalia (1993) and Haiti (1995), and as officer in charge of the Field Operations Branch, Marine Corps History and Museum Division. He earned a Bachelor of Arts degree in history and psychology in 1972, a Master of Arts degree in 1978, and a Doctorate in history in 1985. Dr. Cureton is the author of numerous publications in the field of military history.

BOOKS FROM NORTH CAPE PUBLICATIONS®, INC.

The books in the "For Collectors Only®" and "A Shooter's and Collector's Guide" series are designed to provide the firearms collector with an accurate record of the markings, dimensions and finish found on an original firearm as it was shipped from the factory. As changes to any and all parts are listed by serial number range, the collector can quickly assess not only whether or not the overall firearm is correct as issued, but whether or not each and every part is original for the period of the particular firearm's production. "For Collectors Only" and "A Shooter's and Collector's Guide" books make each collector and shooter an "expert."

FOR COLLECTORS ONLY® SERIES

Serbian and Yugoslav Mausers, by Branko Bogdanovic ($19.95). When Serbia won her independence from Turkey in 1878, the first responsibility of the new government of the small nation hemmed in by predatory powers on all sides was to develop a well-trained and disciplined military force and arm it with the best weapons then available. The Model 1871 Mauser, perhaps the best breech-loading, bolt-action rifle available at the time, was the first choice. And so began a 125-year partnership with the Gebrueder Wilhelm and Paul Mauser Company that endured two world wars, two occupations and a Cold War, and that endures to this day.

It was Serbian gunsmiths who developed the famous "ring of steel" that enclosed the cartridge head and helped make the Model 1898 Mauser rifle a world standard in bolt-action, magazine rifles. As the years passed, Serbia/Yugoslavia adopted the latest Mauser models for their military forces until in 1956, the bolt-action rifle was discarded in favor of the semiautomatic, and then the automatic assault rifle. But the Mauser continues to be manufactured today as a military and police sniper rifle and as a sporting rifle for hunting and target shooting.

Thousands of Yugoslav Mauser rifles have been imported into North America in the last two decades and Mr. Bogdanovic's book will help the collector and shooter determine which model he or she has, and its antecedents. Every Mauser that found its way into military service in Serbia/Yugoslavia is listed and described.

Swiss Magazine Loading Rifles, 1869 to 1958, by Joe Poyer ($19.95). The Swiss were the first to adopt a repeating rifle as general issue to all troops in 1869. The rifle was the Vetterli, a clever blend of Swiss and American engineering. In 1889, the Swiss adopted a small-bore rifle with a straight pull bolt and a box magazine, the Schmidt-Rubin, that somewhat resembled that developed around the same time for the British Lee-Enfield rifles. The design was so successful, that with relatively minor changes and upgrades, it remained in service until 1958 when it was replaced by a semiautomatic rifle. As with all the books in the "For Collectors Only®" series, there is a complete part-by-part description for both the Vetterli and Schmidt-Rubin rifles in all their variations by serial number range, plus a history of their development and use, their cleaning, maintenance and how to shoot them safely and accurately.

The American Krag Rifle and Carbine, by Joe Poyer, edited by Craig Riesch ($19.95). A new look on a part-by-part basis at the first magazine repeating service arm adopted for general service in American military history. It was the arm first adopted for smokeless powder and it required new manufacturing techniques and processes to be developed for its production at Springfield Armory. The Krag was an outstanding weapon that helped define the course of American arms development over the next fifty years. In this new text, the Krag is redefined in terms of its development. Old shibboleths, mischaracterizations and misinterpretations are laid to rest and a true picture of this amazingly collectible rifle and carbine emerges. The author has also devised a monthly serial number chart from production, quarterly and annual reports from Springfield Armory and the Chief of Ordnance to the Secretary of War.

The Model 1903 Springfield Rifle and Its Variations (2nd edition, revised and expanded), by Joe Poyer ($22.95). Includes every model of the Model 1903 from the ramrod bayonet to the Model 1903A4 Sniper rifle. Every part description includes changes by serial number range, markings and finish. Every model is described and identified. Abundant color and black-and-white photos and line drawings of parts to show details precisely. 480 pages.

The .45-70 Springfield (3rd edition, revised and expanded), by Joe Poyer and Craig Riesch ($16.95), covers the entire range of .45-caliber "trapdoor" Springfield arms, the gun that really won the West. "Virtually a mini-encyclopedia . . . this reference piece is a must," Phil Spangenberger, *Guns & Ammo*.

U.S. Winchester Trench and Riot Guns and Other U.S. Combat Shotguns (2nd edition, revised), by Joe Poyer ($16.95). Describes the elusive and little-known "Trench Shotgun" and all other combat shotguns used by U.S. military forces. "U.S. military Models 97 and 12 Trench and Riot Guns, their parts, markings [and] dimensions [are examined] in great detail . . . a basic source of information for collectors," C.R. Suydam, *Gun Report*.

The U.S. M1 Carbine: Wartime Production (4th edition revised), by Craig Riesch ($16.95), describes the four models of M1 Carbines from all ten manufacturers. Complete with codes for every part by serial number range. "The format makes it extremely easy to use. The book is a handy reference for beginning or experienced collectors," Bruce Canfield, Author of *M1 Garand and M1 Carbine*.

The M1 Garand, 1936 to 1957 (4th edition, revised and expanded), by Joe Poyer and Craig Riesch ($19.95). "The book covers such important identification factors as manufacturer's markings, proof marks, final acceptance cartouches stampings, heat treatment lot numbers . . . there are detailed breakdowns of . . . every part . . . in minute detail. This 263 page . . . volume is easy to read and full of identification tables, parts diagrams and other crucial graphics that aid in determining the originality of your M1 and/or its component parts," Phil Spangenberger, *Guns & Ammo*.

Winchester Lever Action Repeating Firearms, by Arthur Pirkle
    Volume 1, The Models of 1866, 1873 & 1876 ($19.95)
    Volume 2, The Models of 1886 and 1892 ($19.95)
    Volume 3, The Models of 1894 and 1895 ($19.95)
These famous lever action repeaters are completely analyzed part-by-part by serial number range in this first new book on these fine weapons in twenty years. ". . . book is truly for the serious collector . . . Mr. Pirkle's scholarship is excellent and his presentation of the information . . . is to be commended," H.G.H., *Man at Arms*.

The SKS Carbine (3rd revised and expanded edition), by Steve Kehaya and Joe Poyer ($16.95). The SKS Carbine "is profusely illustrated, articulately researched and covers all aspects of its development as well as . . . other combat guns used by the USSR and other Communist bloc nations. Each component . . . from stock to bayonet lug, or lack thereof, is covered along with maintenance procedures . . . because of Kehaya's and Poyer's book, I have become the leading expert in West Texas on [the SKS]," Glen Voorhees, Jr., *Gun Week*.

British Enfield Rifles, by Charles R. Stratton
> **Volume 1, SMLE (No. 1) Mk I and Mk III** ($16.95)

"Stratton . . . does an admirable job of . . . making sense of . . . a seemingly hopeless array of marks and models and markings and apparently endless varieties of configurations and conversions . . . this is a book that any collector of SMLE rifles will want," Alan Petrillo, *The Enfield Collector's Digest*.

> **Volume 2, The Lee-Enfield No. 4 and No. 5 Rifles** ($16.95)

In Volume 2, "Skip" Stratton provides a concise but extremely thorough analysis of the famed British World War II rifle, the No. 4 Enfield, and the No. 5 Rifle, better known as the "Jungle Carbine." It's all here, markings, codes, parts, manufacturers and history of development and use.

> **Volume 4, The Pattern 1914 and U.S. Model 1917 Rifles** ($16.95)

In Volume 4, the author describes the events that led to the development of the British Pattern 1914 Enfield and its twin, the U.S Model 1917 Enfield rifle. The M1917 was produced in and used on the Western front in far greater numbers than was the M1903 Springfield. Skip Stratton provides not only the usual part-by-part analysis of both rifles to show how the M1917 evolved from the Pattern 1914, but provides a cross-check of which parts are interchangeable. Included are the sniper and Pedersen Device variants.

The Mosin-Nagant Rifle (3rd revised and expanded edition), by Terence W. Lapin ($19.95). For some reason, in the more than 100 years that the Mosin-Nagant rifle has been in service around the world, not a single book has been written in English about this fine rifle. Now, just as interest in the Mosin-Nagant is exploding, Terence W. Lapin has written a comprehensive volume that covers all aspects and models from the Imperial Russian rifles to the Finnish, American, Polish, Chinese, Romanian and North Korean variations. His book has set a standard that future authors will find very difficult to best. Included are part-by-part descriptions of all makers, Russian, Chinese, American, Polish, Romanian, etc. Also includes all variants such as carbines and sniper rifles from all countries.

The Swedish Mauser Rifles, by Steve Kehaya and Joe Poyer ($19.95). The Swedish Mauser rifle is perhaps the finest of all military rifles manufactured in the late 19th and early 20th centuries. A complete history of the development and use of the Swedish Mauser rifles is provided as well as a part-by-part description of each component. All 24 models are described and a complete description of the sniper rifles and their telescopic sights is included. All markings, codes, regimental and other military markings are charted and explained. A thorough and concise explanation of the Swedish Mauser rifle, both civilian and military.

A Shooter's and Collector's Guide Series

The AK-47 and AK-74 Kalashnikov Rifles and Their Variations, by Joe Poyer ($22.95). The AK-47 and its small-caliber replacement, the AK-74, symbolize for Americans the now-defunct Soviet empire and its support for wars of "national liberation." Author Joe Poyer has examined and described the Kalashnikov rifle on a part-by-part basis, pointing out the differences between the various types of receivers and other parts, as well as the differences between the AK and AKM models. A detailed survey of all models of the Kalashnikov rifle from the AK-47 to the AK-108 is included as are descriptions of those Kalashnikov rifles manufactured by various countries from China to Switzerland. Accessories issued to the soldier from bayonets to web gear are included. Instructions on shooting, selecting telescopic sights, ammunition and troubleshooting round out the book.

The M16/AR15 Rifle (2nd edition, revised and expanded), by Joe Poyer ($19.95). The M16 has been in service longer than any other rifle in the history of the United States military. Its civilian counterpart, the AR15, has recently replaced the M14 as the national match service rifle. This 140-page, profusely illustrated, large-format book examines the development, history and current and future use of the M16/AR15. It describes in detail all civilian AR15 rifles from more than a dozen different manufacturers and takes the reader step-by-step through the process of accurizing the AR15 into an extremely accurate target rifle. Ammunition, both military and civilian, is discussed and detailed assembly/disassembly and troubleshooting instructions are included.

The M14-Type Rifle (2nd edition), by Joe Poyer ($14.95). A study of the U.S. Army's last and short-lived .30-caliber battle rifle which became a popular military sniper and civilian high-power match rifle. A detailed look at the National Match M14 rifle, the M21 sniper rifle and the currently available civilian semiautomatic match rifles, receivers, parts and accessories, including the Chinese M14s. A guide to custom-building a service-type rifle or a match-grade, precision rifle. Includes a list of manufacturers and parts suppliers, plus the BATFE regulations that allow a shooter to build a legal look-alike M14-type rifle.

The SAFN-49 Battle Rifle, by Joe Poyer ($14.95). The SAFN-49, the predecessor of the Free World's battle rifle, the FAL, has long been neglected by arms historians and writers, but not by collectors. Developed in the 1930s at the same time as the M1 Garand and the SVT38/40, the SAFN-49 did not reach production, because of the Nazi invasion of Belgium, until after World War II. This study of the SAFN-49 provides a part-by-part examination of the four calibers in which the rifle was made. Also, contains a thorough discussion of the SAFN-49 Sniper Rifle and its telescopic sights, plus maintenance, assembly/disassembly, accurizing, restoration and shooting. A new exploded view and section view are included. The rifle's development and military use are also explained in detail.

COLLECTOR'S GUIDE TO MILITARY UNIFORMS

The "Collector's Guide to Military Uniforms" endeavors to do for the military uniform collector what the "For Collectors Only®" series does for the firearms collector. Books in this series are carefully researched using original sources; they are heavily illustrated with line drawings and photographs, both period and contemporary, to provide a clear picture of development and use. Where uniforms and accouterments have been reproduced, comparisons between original and reproduction pieces are included so that the collector and historian can differentiate the two.

Campaign Clothing: Field Uniforms of the Indian War Army
    Volume 1, 1866–1871 ($12.95)
    Volume 2, 1872–1886 ($14.95)
Lee A. Rutledge has produced a unique perspective on the uniforms of the Army of the United States during the late Indian War period follow-

ing the Civil War. He discusses what the soldier really wore when on campaign. No white hats and yellow bandanas here.

A Guide Book to U.S. Army Dress Helmets, 1872–1904, by Mark Kasal and Don Moore ($16.95).
From 1872 to 1904, the men and officers of the U.S. Army wore a fancy, plumed or spiked helmet on all dress occasions. As ubiquitous as they were in the late 19th century, they are extremely scarce today. Kasal and Moore have written a step-by-step, part-by-part analysis of both the Models 1872 and 1881 dress helmets and their history and use. Profusely illustrated with black-and-white and color photographs of actual helmets.

All of the above books can be obtained directly from **North Cape Publications®, Inc., P.O. Box 1027, Tustin, CA 92781** or by calling Toll Free 1-800 745-9714. Orders only to the toll-free number, please. For information, call 714 832-3621. Orders may also be placed by Fax (714 832-5302) or via e-mail to ncape@ix.netcom.com. CA residents add 7.75% sales tax. Postage is currently $3.95 for 1-2 books, $5.95 for 3-4 books, $8.95 for 5-12 books. Call, fax or e-mail for UPS and Federal Express rates, for postage on quantities of 13 or more books, and for foreign postage rates.

Also, visit our Internet Website at http://www.northcapepubs.com. Our complete, up-to-date book list can always be found there. Also check out our linked Online Magazine for the latest in firearms-related, magazine-quality articles and excerpts from our books.